Lecture Notes in Control and Information Sciences

Edited by A. V. Balakrishnan and M. Thoma

Vol. 1: Distributed Parameter Systems: Modelling and Identification
Proceedings of the IFIP Working Conference, Rome, Italy, June 21–26, 1976
Edited by A. Ruberti
V, 458 pages. 1978

Vol. 2: New Trends in Systems Analysis
International Symposium, Versailles, December 13–17, 1976
Edited by A. Bensoussan and J. L. Lions
VII, 759 pages. 1977

Vol. 3: Differential Games and Applications
Proceedings of a Workshop, Enschede, Netherlands, March 16–25, 1977
Edited by P. Hagedorn, H. W. Knobloch, and G. J. Olsder
XII, 236 pages. 1977

Vol. 4: M. A. Crane, A. J. Lemoine
An Introduction to the Regenerative Method for Simulation Analysis
VII, 111 pages. 1977

Vol. 5: David J. Clements, Brian D. O. Anderson
Singular Optimal Control: The Linear Quadratic Problem
V, 93 pages. 1978

Vol. 6: Optimization Techniques
Proceedings of the 8th IFIP Conference on Optimization Techniques, Würzburg, September 5–9, 1977
Part 1
Edited by J. Stoer
XIII, 528 pages. 1978

Vol. 7: Optimization Techniques
Proceedings of the 8th IFIP Conference on Optimization Techniques, Würzburg, September 5–9, 1977
Part 2
Edited by J. Stoer
XIII, 512 pages. 1978

Vol. 8: R. F. Curtain, A. J. Pritchard
Infinite Dimensional Linear Systems Theory
VII, 298 pages. 1978

Vol. 9: Y. M. El-Fattah, C. Foulard
Learning Systems:
Decision, Simulation, and Control
VII, 119 pages. 1978

Vol. 10: J. M. Maciejowski
The Modelling of Systems with Small Observation Sets
VII, 241 pages. 1978

Vol. 11: Y. Sewaragi, T. Soeda, S. Omatu
Modelling, Estimation, and Their Applications for Distributed Parameter Systems
VI, 269 pages. 1978

Vol. 12: I. Postlethwaite, A. G. J. McFarlane
A Complex Variable Approach to the Analysis of Linear Multivariable Feedback Systems
IV, 177 pages. 1979

Vol. 13: E. D. Sontag
Polynomial Response Maps
VIII, 168 pages. 1979

Vol. 14: International Symposium on Systems Optimization and Analysis
Rocquentcourt, December 11–13, 1978;
IRIA LABORIA
Edited by A. Bensoussan and J. Lions
VIII, 332 pages. 1979

Vol. 15: Semi-Infinite Programming
Proceedings of a Workshop, Bad Honnef, August 30 – September 1, 1978
V, 180 pages. 1979

Vol. 16: Stochastic Control Theory and Stochastic Differential Systems
Proceedings of a Workshop of the „Sonderforschungsbereich 72 der Deutschen Forschungsgemeinschaft an der Universität Bonn"
which took place in January 1979 at Bad Honnef
VIII, 615 pages. 1979

Vol. 17: O. I. Franksen, P. Falster, F. J. Evans
Qualitative Aspects of Large Scale Systems
Developing Design Rules Using APL
XII, 119 pages. 1979

Vol. 18: Modelling and Optimization of Complex Systems
Proceedings of the IFIP-TC 7 Working Conference
Novosibirsk, USSR, 3–9 July, 1978
Edited by G. I. Marchuk
VI, 293 pages. 1979

Vol. 19: Global and Large Scale System Models
Proceedings of the Center for Advanced Studies (CAS)
International Summer Seminar
Dubrovnik, Yugoslavia, August 21–26, 1978
Edited by B. Lazarevic
VIII, 232 pages. 1979

Vol. 20: B. Egardt
Stability of Adaptive Controllers
V, 158 pages. 1979

Vol. 21: Martin B. Zarrop
Optimal Experiment Design for Dynamic System Identification
X, 197 pages. 1979

Vol. 26: D. L. Iglehart, G. S. Shedler
Regenerative Simulation of Response Times in Networks of Queues
XII, 204 pages. 1980

Lecture Notes in Control and Information Sciences

Edited by A. V. Balakrishnan and M. Thoma

26

D. L. Iglehart · G. S. Shedler

Regenerative Simulation of Response Times in Networks of Queues

Springer-Verlag
Berlin Heidelberg GmbH 1980

Authors
Donald L. Iglehart
Department of Operations Research
Stanford University
Stanford, California 94305
and
Gerald S. Shedler
IBM Research Laboratory
San Jose, California 95193

ISBN 978-3-540-09942-0 ISBN 978-3-540-39151-7 (eBook)
DOI 10.1007/978-3-540-39151-7

Originally published by Springer-Verlag Berlin Heidelberg New York in 1980

2060/3020-543210

PREFACE

Discrete event digital simulation of stochastic models has been one of the most important practical tools of systems analysis for well over twenty years. The complexity of models for most real systems is such that we are unable to study them analytically. Computer simulation is our only alternative and we must seek theoretically sound and computationally efficient methods for carrying out the simulation. While a great deal of effort has been devoted to the development of simulation programming languages and programs, relatively little has been done to develop theoretical foundations to justify the estimation methods implemented in the simulations. Typically, simulation program packages compute by rote and report a large variety of point estimates for various characteristics of the model being simulated. Seldom do these reports indicate the variability or statistical precision of the point estimates.

This monograph deals with probabilistic and statistical methods for discrete event simulation of networks of queues. The emphasis is on the use of underlying stochastic structure for the design of simulation experiments and the analysis of simulation output. We focus on recently developed methods for estimation of general characteristics of "passage times" in closed networks of queues. Informally, a passage time is the time for a job to traverse a portion of a network. Such quantities are important in computer and communication system models where they represent job response times. In this context, expected values as well as other characteristics of response times are of interest.

The presentation is self contained. We have attempted to make this material accessible to simulation practitioners as well as to students and researchers interested in the methodology of discrete event simulation. For this reason, we have provided a number of examples and have separated the exposition of the estimation procedures from the derivations. Some knowledge of elementary probability theory, statistics, and stochastic models is sufficient to understand the estimation procedures and the examples. The derivations use results often contained in first year graduate courses in stochastic processes. These sections can be omitted by the reader interested primarily in application of the procedures.

We have benefitted from the comments of Peter Glynn, Lily Jow, Austin Lemoine, and Gerard Scallan who read an earlier draft of the monograph. It is a pleasure to acknowledge our gratitude to Julie Countryman and Blanca Gallegos of the Manuscript Processing Center at the IBM San Jose Research Laboratory for their expert typing of a difficult manuscript. We are also indebted to Jean Chen and Jack DeLany of Central Scientific Services at IBM Research for their careful preparation of the figures.

We are both grateful to the National Science Foundation for support under Grant MCS-7909139. In addition, one of us (D.L.I.) gratefully acknowledges partial support under National Science Foundation Grant MCS-7523607 and Office of Naval Research Contract N00014-76-C-0578 (NR 042-343).

Donald L. Iglehart
Gerald S. Shedler

Stanford University
Stanford, California

IBM Research Laboratory
San Jose, California

November, 1979

TABLE OF CONTENTS

LIST OF FIGURES

LIST OF TABLES

1.0. INTRODUCTION

Networks of queues occur frequently in diverse applications. In particular, they are widely used in studies of computer and communication system performance as models for the interactions among system resources. This monograph deals with mathematical and statistical methods for discrete event simulation of networks of queues. The emphasis is on methods for the estimation of general characteristics of "passage times" in closed networks. Informally, a passage time is the time for a job to traverse a portion of a network. Such quantities, calculated as random sums of queueing times, are important in computer and communication system models where they represent job response times.

By simulation we mean observation of the behavior of a stochastic system of interest by artificial sampling on a digital computer. With discrete event simulation, stochastic changes of the system state occur only at a set of increasing time points. Simulation is a tool which can be used to study complex stochastic systems when analytic and/or numerical techniques do not suffice; in connection with the study of complex networks of queues encountered in applications, this is often the case.

When simulating, we experiment with a stochastic system and observe its behavior. During the simulation we measure certain quantities in the system and, using statistical techniques, draw inferences about characteristics of well defined random variables. The most obvious methodological advantage of simulation is that in principle it is applicable to stochastic systems of arbitrary complexity. In practice, however, it is often a decidedly nontrivial matter to obtain from a simulation information which is both useful and accurate, and to obtain it in an efficient manner. These difficulties arise primarily from the inherent variability in a stochastic system, and it is necessary to seek theoretically sound and computationally efficient methods for carrying out the simulation. Apart from implementation considerations, important concerns for simulation relate to efficient methods for generation of realizations (sample paths) of the stochastic system under study, the design of simulation experiments, and the analysis of simulation output. It is fundamental for simulation, since results are based on observation of a stochastic system, that some assessment of the precision of results be provided.

Assessing the statistical precision of a point estimate requires careful design of the simulation experiments and analysis of the simulation output. In general, the desired statistical precision takes the form of a confidence interval for the quantity of interest. Among the issues the simulator must face are the initial conditions for the system being simulated, the length of the simulation run, the number of replications of the experiments, and the length of the confidence interval. Over the last five years, there has been increased attention paid to these issues, and a theory of simulation analysis (the regenerative method) has been developed which, when applicable, provides some measure of statistical precision. The regenerative method, which is based on limit theorems developed for regenerative stochastic processes, plays a key role in our discussion of simulation methods for passage times in networks of queues.

Under the usual queueing-theoretic (independent and identically distributed service and interarrival time) assumptions, analyses based on a "numbers-in-queue" and "stages-of-service" state vector can be carried out. Typically it is necessary to assume that all service and interarrival time distributions are exponential or have a Cox-phase (exponential stage) representation. Under these assumptions, expressions suitable for numerical evaluation are obtainable for queue length distributions. Other measures of system performance (calculated as random sums of queueing times) involve the times, here called passage times, for a job to traverse a portion of the network. Often when such quantities arise in computer and communication system models, they represent job response times. In this context, expected values as well as other characteristics of passage times (e.g., percentiles) are of interest. The analyses based on the numbers-in-queue, stages-of-service state vector yield expected values for passage times, but do not yield other passage time characteristics of interest. Moreover, alternative analyses to provide these measures of the variability of system response are in general not available, and it is necessary to resort to simulation. Although the usual process of numbers-in-queue and stages-of-service is a regenerative process (in fact a Markov chain) under the probabilistic assumptions that we make here, the regenerative method cannot be applied directly to this process to estimate general passage time characteristics. This is essentially because passage times are not totally contained within cycles of the numbers-in-queue, stages-of-service process.

The organization of the presentation is as follows. This initial section provides some motivation for study of simulation methods for passage times in networks of queues, a brief overview of some of the methodological considerations for simulation, and a summary of the discussion which appears in subsequent sections.

The estimation methods developed here for passage times in networks of queues use the regenerative method for analysis of simulation output. Based on a single simulation run, these methods provide (strongly consistent) point estimates and (asymptotically) valid confidence intervals for general characteristics of limiting passage times. Section 2 provides a review of the regenerative method. The section contains a brief discussion of the underlying theory of regenerative stochastic processes along with some examples of regenerative processes in networks of queues.

Section 3 provides a specification of the basic class of closed networks of queues with which we deal, and the probabilistic assumptions therein. Initially, we restrict attention to networks with stochastically identical jobs and give a state vector definition based on a linear job stack. The section also contains the formal definition of passage time in a network of queues.

The notion of a distinguished "marked" job is fundamental to the method for estimation of passage time characteristics described and developed in Section 4. The approach is to consider a Markov renewal process arising from a continuous time Markov chain defined by the usual numbers-in-queue, stages-of-service state vector augmented by information sufficient to track the marked job. We arbitrarily select a job to serve as the marked job and measure its passage times during the simulation. They key steps in the derivation of this marked job method are identification of an appropriate regenerative process in discrete time and development of a ratio formula from which point estimates and confidence intervals can be obtained for quantities associated with the limiting passage time.

In Section 5 we consider application of the marked job method to two particular closed networks of queues, and display some numerical results. The first example is a relatively simple network. Despite the apparent structural simplicity of this network, it exhibits the essence of the passage time simulation problem. The second and more complex network arises as a model for a computer data base management system. This model illustrates the representation of complex congestion phenomena in the framework of Section 3.

The extension of the marked job method to certain finite capacity open networks of queues is the subject of Section 6. Particular stochastic point processes associated with a Markov renewal process generate arrivals to the networks, and there are two formulations of the finite capacity constraint. The network structure we permit is essentially the same as that described in Section 3 except that here the networks are open. To estimate passage times in these networks, we track an appropriate sequence of typical jobs, based on the idea of a marked job. These are to be typical jobs in the sense that the sequence of passage times for the marked jobs should converge in distribution to the same random variable as do the passage times for all the jobs. It is necessary to take some care to ensure that this is the case.

Slightly restricting the definition of passage time given in Section 3, we develop in the next section a new and somewhat simpler stochastic setting than that of Section 4 for the marked job method. Here the starts of passage times for the marked job are the successive entrances of a particular continuous time Markov chain to a fixed set of states, and the terminations of such passage times are the successive entrances to another fixed set of states. This fomulation, in terms of hitting times to fixed sets of states, is the basis for the passage time simulation method developed in the next section.

The marked job method of Section 5 is applicable to passage times in the general sense, i.e., whether or not the passage time is a complete circuit. For those passage times (termed "response times") which are complete circuits in a closed network, simulation using a marked job appears to be the only method available for obtaining confidence intervals from a single simulation run. It is inherent in the marked job method that only the passage times observed for the marked job enter into the construction of point and interval estimates, and we would expect some loss of efficiency as the price for obtaining confidence intervals. In Section 8, we concentrate on passage times through a subnetwork of a given network of queues, and develop the <u>decomposition method</u>. With this estimation method, passage times observed for all the jobs during a single simulation run enter into the construction of point and interval estimates. The basis for this method is the observation that the successive entrances to a fixed set of states of an appropriate continuous time stochastic process serve to decompose the sequence of passage times for all the jobs into independent and identically distributed blocks.

Section 9 deals with the statistical efficiency of the marked job and decomposition methods. We consider the calculation of theoretical values for variance constants entering into central limit theorems used to obtain confidence intervals for mean passage times. The results of this section provide a firm basis for comparing the efficiency of the two methods where both apply. They can also be used to give some idea of the efficiency of the marked job method in the case of response times, where it is our only means of obtaining confidence intervals from a single simulation run.

The estimation of passage times in closed networks of queues with multiple job types is the subject of Section 10. Here the type of a job may influence its routing through the network as well as its service requirements at each center. Using the stochastic setting of Section 7 for the marked job method, we mark one job of each type. By tracking these marked jobs through the network, we are able to obtain point and interval estimates for a variety of measures of the variability of system response over the several job types.

The final section considers some aspects of uniform and nonuniform random number generation pertinent to implementation of a passage time simulation. We also discuss the use of random number streams and the generation of state vector processes.

2.0. SIMULATION OF REGENERATIVE PROCESSES

When the output of a discrete event digital simulation is a stochastic process $\underset{\sim}{X}=\{X(t):t\geq 0\}$ that approaches a "steady state" which is of interest, it may be possible to characterize the stochastic structure of the process and to use this structure in carrying out the simulation. In such cases, mathematical results on the stochastic structure of the process $\underset{\sim}{X}$ form the basis for both the design of simulation experiments and the analysis of simulation output. For particular stochastic processes known as regenerative processes, Fishman (1973) and Crane and Iglehart (1975a) have provided a theory of simulation analysis called the regenerative method. This theory has been developed in subsequent papers including Crane and Iglehart (1975b), Iglehart (1975), (1976), Hordijk, Iglehart and Schassberger (1976), and Lavenberg and Sauer (1977); see Crane and Lemoine (1977) and Iglehart (1978) for an introduction to and a detailed review of the regenerative method. Regenerative simulation underlies the estimation methods described in subsequent sections for passage times in networks of queues. In this section we discuss regenerative processes and review the regenerative method.

Heuristically, a regenerative stochastic process is a process having the property that there exist random time points at which the process probabilistically restarts. Typically, these time points at which the process probabilistically starts afresh, referred to as regeneration points or regeneration times, are returns to a fixed state of the process. The essential idea of a regenerative process is that between any two successive regeneration points, the evolution of the process is a probabilistic replica of the process between any other such pair of regeneration points. For many stochastic models in which it is possible to identify a sequence of regeneration points, this discovery is a key to an analytic/numerical solution which yields expressions for quantities of interest. It is often the case, however, that even though a sequence of regeneration points exists, it is nevertheless not possible to obtain an analytic/numerical solution because of severe computational difficulties; we would then consider simulation, and it is this situation with which we deal here.

The regenerative process structure, in the presence of certain regularity conditions, guarantees the existence of a "steady state" for the process, i.e., that there exists a random variable X such that

$$\lim_{t\to\infty} P\{X(t)\leq x\} = P\{X\leq x\} ,$$

for all x at which the right hand side of this equation is continuous. This type of convergence (in distribution) is known as <u>weak convergence</u> and we denote it by "$X(t)\Rightarrow X$ as $t\to\infty$." Furthermore, the regenerative structure ensures that the "steady state" X of the process is determined (as a ratio of expected values) by the behavior of the process between two successive regeneration points. There is an important implication of these mathematical results for the simulation of regeneration processes. A strongly consistent point estimate and asymptotically valid confidence interval for the expected value of a general (measurable) function of the steady state X can be obtained by observation (in cycles of random length defined by the regeneration points) of a finite portion of a single realization of the process $\underset{\sim}{X}$. When simulating (say, for a fixed number of cycles), we measure appropriate quantities defined totally within the individual cycles and compute sample means over the cycles.

Where applicable, the regenerative method has great appeal because it provides both point and interval estimates having desirable properties. There are, however, other considerations. The classical alternative for estimation of "steady state" quantities would entail selecting an initial state for the process, running the simulation for an initial period of time (and discarding this "initial transient"), and then observing the process ("in steady state") for an additional period of time from which point estimates are obtained. In general, no confidence interval is available, nor is there any guidance on the selection of the initial state. Moreover, the determination of appropriate initial and additional periods of time is often nontrivial and likely to require sophisticated statistical techniques. There are similar problems with simulation methods based on multiple replications. With the regenerative method, these difficulties to a large extent are avoidable.

2.1. Definition of Regenerative Process

A regenerative process in continuous time can be defined in terms of the pasting together of so-called "tours"; see Smith (1958) and Miller (1972). The formal definition of a regenerative process that we give is equivalent to these and also to the definition of Cinlar (1975a), p. 298. We require the notion of a renewal process and that of a stopping time for a stochastic process.

A sequence of random variables $\underset{\sim}{T}=\{T_n:n\geq 0\}$ is a <u>renewal process</u> provided that $T_0=0$ and T_n-T_{n-1} $(n\geq 1)$ are independent, identically distributed (i.i.d.) positive random variables. We always assume that $\underset{\sim}{T}$ is persistent, i.e., that $P\{T_n-T_{n-1}<\infty\}=1$. A random variable T taking values in $[0,+\infty)$ is a <u>stopping time</u> for a stochastic process $\underset{\sim}{X}$ provided that for every finite $t\geq 0$, the occurrence or nonoccurrence of the event $\{T\leq t\}$ can be determined from the history $\{X(u):u\leq t\}$ of the process up to time t; see Çinlar (1975a), p. 239 for a discussion of stopping times.

(2.1.1) DEFINITION. The real (possibly vector-valued) stochastic process $\underset{\sim}{X}=\{X(t):t\geq 0\}$ is a regenerative process if

(i) there exists a sequence of stopping times $\underset{\sim}{\beta}=\{\beta_n:n\geq 0\}$ such that $\underset{\sim}{\beta}$ is a renewal process; and

(ii) for every sequence of times $0<t_1<t_2<\ldots<t_m$ $(m\geq 1)$ and $n\geq 0$, the random vectors $\{X(t_1),\ldots,X(t_m)\}$ and $\{X(\beta_n+t_1),\ldots,X(\beta_n+t_m)\}$ have the same distribution, and the processes $\{X(t):t<\beta_n\}$ and $\{X(\beta_n+t):t\geq 0\}$ are independent.

The points of $\underset{\sim}{\beta}$ are the <u>regeneration points</u> for the process $\underset{\sim}{X}$ and we refer to the interval $[\beta_{n-1},\beta_n)$ as the nth <u>cycle</u> of the regenerative process. The definition of a regenerative process in discrete time is similar; see the discussion of recurrent events in Feller (1968), Ch. XIII.

It is straightforward to check that irreducible and positive recurrent (continuous or discrete time) Markov chains and semi-Markov processes having finite (or countable) state space are regenerative processes. It can also be shown that in a single server queueing system in which the sequence of successive interarrival and service times are positive, i.i.d. random vectors (having finite means), the processes of the number of jobs $Q(t)$ in the system at time t, the waiting time W_n of the nth job, and the virtual waiting time $V(t)$ at time t are all regenerative processes, provided that the traffic intensity is less than 1. For the process $\{W_n:n\geq 0\}$, the regeneration points are the indices of jobs having zero waiting time. For the process $\{Q(t):t\geq 0\}$ as well as the process $\{V(t):t\geq 0\}$, the regeneration points are the start of a period during which the server is busy.

In developing the regenerative method, it is necessary to distinguish two cases. For the renewal process $\underset{\sim}{\beta}=\{\beta_n:n\geq 0\}$ in the definition of the regenerative process $\underset{\sim}{X}$, we let $\alpha_n=\beta_n-\beta_{n-1}$ $(n\geq 1)$ and denote by F the distribution function of α_n. The random variable α_1 (or distribution function F) is said to be <u>periodic</u> with period $\lambda>0$ if, with probability one, α_1 assumes values in the set $\{0,\lambda,2\lambda,...,\}$ and λ is the largest such number. If there is no such λ, then α_1 (or F) is said to be <u>aperiodic</u>.

In the aperiodic case, in order for a regenerative process to have a limiting distribution, it is necessary either to impose regularity conditions on the sample paths of the process, or to place restrictions on the distribution function of the time between regeneration points. To be more specific, we first define an appropriate class of distribution functions. Let F_n be the n-fold convolution of the distribution function F, and define $\mathscr{A}$ to be the set of all distribution functions F such that for some $n\geq 1$, F_n has an absolutely continuous component (i.e., has a density function on some interval). It is convenient to write $\alpha_1 \in \mathscr{A}$ when the distribution function F of α_1 is an element of $\mathscr{A}$. We would expect the aperiodic distributions F arising in applications to be elements of the set $\mathscr{A}$. With respect to appropriate regularity conditions on the sample paths of the process, we restrict attention to processes $\underset{\sim}{X}=\{X(t):t\geq 0\}$ having right continuous sample paths and limits from the left, i.e., for $t\geq 0$,

$$X(t) = \lim_{u \downarrow t} X(u)$$

and for all $t>0$,

$$X(t-) = \lim_{u \uparrow t} X(u) \text{ exists },$$

with probability one. For such a k-dimensional stochastic process we write $\underset{\sim}{X}\in D_k[0,\infty)$. With these definitions, we can state the basic limit theorem for regenerative processes.

(2.1.2) THEOREM. Assume that α_1 is aperiodic with $E\{\alpha_1\}<\infty$. If either $\underset{\sim}{X}\in D_k[0,\infty)$ or $\alpha_1 \in \mathscr{A}$, then $X(t) \Rightarrow X$ as $t\to\infty$.

There is a corresponding result for the periodic case. The proof of this theorem (Miller (1972)) involves an application of the key renewal theorem and is somewhat involved technically.

Now suppose that α_1 is aperiodic and that for a real-valued (measurable) function f having domain E, the state space of the process $\underset{\sim}{X}$, the quantity of interest is

$$r(f) = E\{f(X)\} .$$

When the state space E is not discrete, we must also assume that the set D(f) of discontinuities of f is such that $P\{X \in D(f)\}=0$; if E is discrete, we can choose f arbitrarily. We always assume that with probability one the process $f(\underset{\sim}{X})$ is integrable over a finite interval, and for $n \geq 1$ define

$$Y_n(f) = \int_{\beta_{n-1}}^{\beta_n} f(X(u))du .$$

In the case of a regenerative process in discrete time, for $n \geq 1$ we define

$$Y_n(f) = \sum_{k=\beta_{n-1}}^{\beta_n - 1} f(X_k) .$$

Theorems (2.1.3) and (2.1.4) deal with the structure of regenerative processes. These results form the basis for the regenerative method.

(2.1.3) THEOREM. The sequence $\{(Y_n(f), \alpha_n):n \geq 1\}$ consists of independently and identically distributed random vectors.

This follows directly from the definition of a regenerative process. The next result (cf. Crane and Iglehart (1975a), Appendix) provides a ratio formula for the quantity r(f).

(2.1.4) THEOREM. Assume that α_1 is aperiodic with $E\{\alpha_1\}<\infty$, and that $E\{|f(X)|\}<\infty$. If either $f(\underset{\sim}{X}) \in D_1[0,\infty)$ or $\alpha_1 \in \mathcal{A}$, then

$$E\{f(X)\} = E\{Y_1(f)\}/E\{\alpha_1\} .$$

There is an analogous ratio formula when α_1 is periodic. Note that if the state space E is discrete, the condition $f(\underset{\sim}{X})\in D_1[0,\infty)$ always holds.

We now indicate how to obtain a point estimate and confidence interval for the quantity $r(f)=E\{f(X)\}$ from a sample path of the process $\underset{\sim}{X}$. In this discussion, we assume that the regenerative process $\underset{\sim}{X}$ and the function f are such that the ratio formula for r(f) holds. For $k\geq 1$, let

$$Z_k(f) = Y_k(f) - r(f)\alpha_k \tag{2.1.5}$$

and denote the variance of $Z_1(f)$ by $\sigma^2=\text{var}\{Z_1(f)\}$. Note that $\{Z_k(f):k\geq 1\}$ consists of i.i.d. random variables, that $Z_k(f)$ is completely determined by the kth cycle of the regenerative process $\underset{\sim}{X}$, and that $E\{Z_k(f)\}=0$. Writing

$$E\{(Z_1(f))^2\} = E\{[(Y_1(f)-E\{Y_1(f)\})-r(f)(\alpha_1-E\{\alpha_1\})]^2\} ,$$

it follows that

$$\sigma^2 = \text{var}\{Y_1(f)\} - 2r(f)\ \text{cov}\{Y_1(f),\alpha_1\} + (r(f))^2\ \text{var}\{\alpha_1\} . \tag{2.1.6}$$

We require the further assumption that $0<\sigma^2<\infty$. The case $\sigma^2=0$ is degenerate, and $\sigma^2<\infty$ for most finite state processes. (In some queueing systems, however, additional finite higher moment conditions on service and interarrival times are needed to ensure $\sigma^2<\infty$.) For fixed n, we use the quantities $\overline{Y}_n(f)$, $\overline{\alpha}_n$, s_{11}, s_{22} and s_{12} to estimate σ^2; obtained from $(Y_1(f),\alpha_1),\ldots,(Y_n(f),\alpha_n)$, these are the usual unbiased estimators of $E\{Y_1(f)\}$, $E\{\alpha_1\}$, $\text{var}\{Y_1(f)\}$, $\text{var}\{\alpha_1\}$ and $\text{cov}\{Y_1(f),\alpha_1\}$, respectively. As a consequence of the strong law of large numbers for i.i.d. sequences of random variables, as $n\to\infty$ the point estimates

$$\hat{r}_n(f) = \overline{Y}_n(f)/\overline{\alpha}_n$$

and

$$s_n = \left\{s_{11}-2\hat{r}_n(f)s_{12}+\left(\hat{r}_n(f)\right)^2 s_{22}\right\}^{1/2}$$

converge with probability one to $r(f)$ and σ, respectively; thus, by definition $\hat{r}_n(f)$ and s_n are strongly consistent estimates.

The basis for the construction of an asymptotically ($n\to\infty$) valid confidence interval for $r(f)$ is a particular central limit theorem (c.l.t.): if $0<\sigma^2<\infty$, then as $n\to\infty$,

$$n^{1/2}\{\hat{r}_n(f)-r(f)\}/[\sigma/E\{\alpha_1\}] \Rightarrow N(0,1) \ , \tag{2.1.7}$$

where $N(0,1)$ is the standardized (mean 0, variance 1) normal random variable.

To derive this result and similar c.l.t.'s later, we need two lemmas on weak convergence. Let $\{X_n:n\geq 1\}$ and $\{Y_n:n\geq 1\}$ be two sequences of random vectors such that X_n and Y_n are defined on a common probability space for all n and have ranges R^k and R^ℓ, where R^k [respectively R^ℓ] is k-dimensional [respectively ℓ-dimensional] Euclidean space. Let c denote a constant-valued random vector. Also, let h map R^k into R^ℓ and denote by $D(h)$ the set of discontinuity points of h.

(2.1.8) LEMMA. If $X_n \Rightarrow X$ and $Y_n \Rightarrow c$, then

$$(X_n, Y_n) \Rightarrow (X, c) .$$

(2.1.9) LEMMA. If $X_n \Rightarrow X$ and either h is continuous or $P\{X \in D(h)\} = 0$, then

$$h(X_n) \Rightarrow h(X) .$$

Lemma (2.1.8) [respectively Lemma (2.1.9)] is a special case of Theorem 4.4 [respectively Theorem 5.1] of Billingsley (1968).

For the proof of Equation (2.1.7), we first note that the standard c.l.t. for i.i.d., mean 0, finite variance random variables implies that

$$\frac{1}{\sigma n^{1/2}} \sum_{k=1}^{n} Z_k(f) \Rightarrow N(0,1) . \qquad (2.1.10)$$

This can be rewritten as

$$n^{1/2}\{\hat{r}_n(f) - r(f)\}/[(\sigma/E\{\alpha_1\})(E\{\alpha_1\}/\bar{\alpha}_n)] \Rightarrow N(0,1) . \qquad (2.1.11)$$

The strong law of large numbers (or even the weak law for that matter) guarantees that

$$(E\{\alpha_1\}/\bar{\alpha}_n) \Rightarrow 1 . \qquad (2.1.12)$$

Lemma (2.1.8) applies to this situation, and hence if we let X_n denote the left-hand side of Equation (2.1.11),

$$(X_n, (E\{\alpha_1\}/\bar{\alpha}_n)) \Rightarrow (N(0,1), 1) .$$

Now apply Lemma (2.1.9) using the continuous mapping $h(x,y) = x \cdot y$ to conclude that

$$X_n \cdot (E\{\alpha_1\}/\bar{\alpha}_n) \Rightarrow N(0,1)\cdot 1 \ . \qquad (2.1.13)$$

Since $N(0,1)\cdot 1$ has the same distribution as $N(0,1)$, Equation (2.1.13) is the same as Equation (2.1.7).

Equation (2.1.7) provides a confidence interval for $r(f)$ but in general, of course, the "standard deviation constant" $\sigma/E\{\alpha_1\}$ is not available and must be estimated. The most straightforward estimate for $\sigma/E\{\alpha_1\}$ is $s_n/\bar{\alpha}_n$, and the strong law of large numbers ensures that

$$(\sigma/s_n) \Rightarrow 1 \ .$$

Then, by the same argument which leads to Equation (2.1.7), we obtain from Equation (2.1.10) the c.l.t.

$$n^{1/2}\{\hat{r}_n(f)-r(f)\}/(s_n/\bar{\alpha}_n) \Rightarrow N(0,1) \ .$$

It follows that for $0<\gamma<1/2$ the interval

$$\hat{I}_n(f) = \left[\hat{r}_n(f)-z_{1-\gamma}s_n/\left(\bar{\alpha}_n n^{1/2}\right),\ \hat{r}_n(f)+z_{1-\gamma}s_n/\left(\bar{\alpha}_n n^{1/2}\right)\right] ,$$

where $z_{1-\gamma}=\Phi^{-1}(1-\gamma)$ and $\Phi(\cdot)$ is the distribution function of the standardized normal random variable, provides an asymptotically valid $100(1-2\gamma)\%$ <u>confidence interval</u> for $r(f)$. This means that

$$\lim_{n\to\infty} P\{r(f)\in\hat{I}_n(f)\} = 1-2\gamma \ ,$$

and thus when simulating, for large n the interval $\hat{I}_n(f)$ (having random end points) surrounds the unknown constant $r(f)$ approximately $100(1-2\gamma)\%$ of the time. Other point and interval estimates which reduce the bias of

$\hat{r}_n(f)$ are available; see Iglehart (1975). Note that the interval $\hat{I}_n(f)$ is symmetric about the point estimate $\hat{r}_n(f)$, and that the half length of the interval is $n^{-1/2}$ times a multiple ($z_{1-\gamma}$) of the estimate of the constant $\sigma/E\{\alpha_1\}$. Thus as n increases, the length of the interval converges to 0 and the midpoint converges to the true value.

The procedure just described for obtaining confidence intervals when simulating a regenerative process is for a fixed number of cycles. For a simulation of fixed run length t, the procedure (Crane and Iglehart (1975a), p. 39) is the same except that statistics are computed only for the (random number of) cycles completed by time t. Confidence intervals are based on the c.l.t., analogous to Equation (2.1.7), that as $t\to\infty$,

$$t^{1/2}\{\hat{r}_{n(t)}(f)-r(f)\}/[\sigma/(E\{\alpha_1\})^{1/2}] \Rightarrow N(0,1) ,$$

where n(t) is the number of cycles completed in (0,t].

Crane and Iglehart (1975a) have also shown that for regenerative processes possessing more than one sequence of regeneration points, with high probability the resulting confidence intervals are of the same length, provided that the length of the simulation run is large. More precisely, for fixed run length t, if I(t) and $I^*(t)$ are the lengths of confidence intervals obtained from two sequences of regeneration points, then

$$\lim_{t\to\infty} I(t)/I^*(t) = 1$$

with probability one.

Finally, suppose that we are interested in estimating r(f) with a 100(1-2γ)% confidence interval whose half length is 100δ% of r(f). Then the number of cycles required is (approximately)

$$n = (z_{1-\gamma}/\delta)^2[\sigma/(r(f)E\{\alpha_1\})]^2 .$$

The first factor $(z_{1-\gamma}/\delta)^2$ is independent of the system being simulated. From the second factor $[\sigma/(r(f)E\{\alpha_1\})]^2$ it is apparent that some systems are inherently more difficult to simulate than others. This quantity provides a good measure of the length of simulation required for a fixed level of precision. An estimate for this quantity obtained from a pilot run can be used to determine the length of the final simulation run.

2.2. Regenerative Processes in Networks of Queues

To illustrate the ideas of Section 2.1, consider the simple closed network of queues of Figure 2.1. A fixed number of jobs, N, circulate in the network from time t=0. Upon completion of α service at center 1, a job joins the tail of the queue in center 2 for β service. After completion of service at center 2, the job joins the tail of the queue in center 1. Neither center 1 nor center 2 service is subject to interruption, and we assume that both queues are served according to a first-come, first-served (FCFS) discipline. We complete the specification of this network of queues by making the following probabilistic assumptions:

(i) successive α service times form a sequence of i.i.d. random variables exponentially distributed (as S_1) and having rate parameter λ_1, i.e., for t≥0,

$$P\{S_1 \le t\} = 1 - \exp(-\lambda_1 t) ;$$

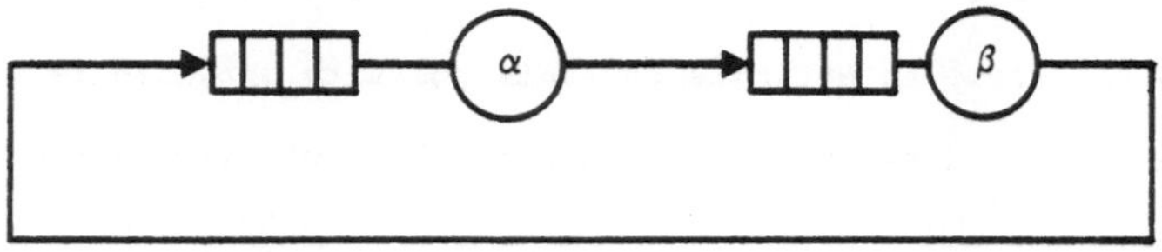

Figure 2.1. Cyclic queues

(ii) successive β service times form a sequence of i.i.d. random variables exponentially distributed (as S_2) and having rate parameter λ_2;

(iii) the sequences in (i) and (ii) are mutually independent.

To study quantities associated with the lengths of the queues in centers 1 and 2, for $t \geq 0$ let $X(t)$ be the number of jobs waiting or in service at center 1 at time t. The stochastic process $\underset{\sim}{X}=\{X(t):t\geq 0\}$ has finite state space $E=\{0,1,...,N\}$, and under assumptions (i), (ii), and (iii) is in fact a continuous time Markov chain. It is easy to check that the Markov chain $\underset{\sim}{X}$ is irreducible and positive recurrent. Thus, this "state vector process" $\underset{\sim}{X}$ for the network of queues is a regenerative process; successive returns to any fixed state $i \in E$ are regeneration points for the process. The regenerative process $\underset{\sim}{X}$ has a "steady state" X, and we can apply the regenerative method to obtain from a single replication point estimates and confidence intervals for quantities of the form $r(f)=E\{f(X)\}$ for some function f having domain E. Suppose for example, that f is the indicator function, $1_{\{1,2,...,N\}}$, of the set $\{1,2,...,N\}$. (The indicator function $1_{\{1,2,...,N\}}(x)$ equals 1 if $x \in \{1,2,...,N\}$ and equals 0 if $x \notin \{1,2,...,N\}$.) Then $r(f)=E\{f(X)\}$ is the steady state utilization of service center 1, i.e., the (limiting) probability that center 1 is busy. If $f(x)=x$, then $r(f)$ is the steady state mean number of jobs waiting or in service at center 1.

Note that if we assume that one of the service time random variables has an exponential distribution, but that the other (say S_2) has a finite mean but otherwise arbitrary distribution, the stochastic process $\underset{\sim}{X}$ is no longer a continuous time Markov chain. The process, however, is still regenerative; the successive entrances of the process $\underset{\sim}{X}$ to a fixed state $i\epsilon\{1,2,\ldots,N\}$ from state i-1 are regeneration points. These time points correspond to completions of service at center 2 after which there are i jobs at center 1.

Now consider the network of queues in Figure 2.2, formulated (Lewis and Shedler (1971)) as a model of system overhead in multiprogrammed computer systems operating under demand paging. The interpretation of this figure differs from that of a conventional diagram for a network of queues in that services are distinguished from the servers which perform them. In Figure 2.2, the circles represent services rather than servers. The model consists of two sequential stages, the α stage and the β stage, in a loop. Two servers, interpreted as a processor and a data transfer unit (IO unit), provide service to a constant number, N, of jobs (programs); each of the jobs goes through both stages in sequence and then returns to the first stage, this process being repeated continuously. Within the α stage, a job receives each of three services α_1, α_2, and α_3, in that order and similarly, within the β stage, a job receives each of three services β_1, β_2, and β_3, in that order. A β_2 service can be provided only by the IO unit, and each of the other services can be provided only by the processor. We assume that the two servers can provide service concurrently, subject to the restriction that the processor cannot provide a β_1 or a β_3 service while the IO unit is providing a β_2 service. In addition, we assume that after having received an α_3 service, a job moves

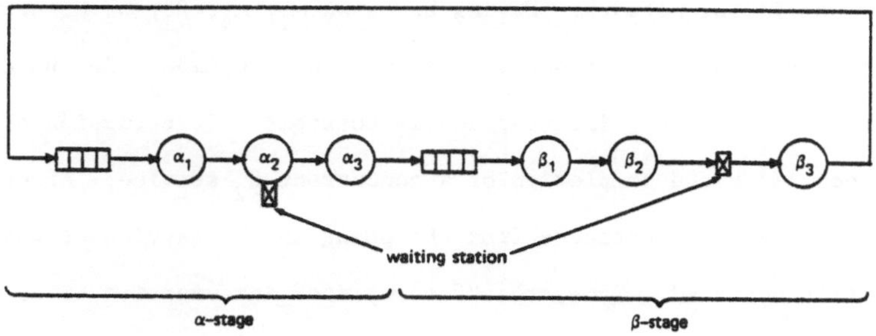

(i) Processor provides α_1, α_2, α_3, β_1 and β_3 services

(ii) I/O unit provides β_2 service

(iii) No β_1, β_3 service by processor during β_2 service by I/O unit

(iv) α_1, α_3, β_1, β_2, β_3 services not interruptable

(v) α_2 service has pre-emptive resume type interruption at completion of β_2 service

Figure 2.2. System overhead model

instantaneously from the α stage to the tail of the queue in the β stage, and after having received a β_3 service, moves instantaneously from the β stage to the tail of the queue in the α stage.

The single processor provides α_1, α_2, α_3, β_1, and β_3 service to the several jobs in the network. Having begun an α_1, α_3, β_1, or β_3 service, the processor completes the service without interruption. In the case of the α_2 service, however, interruption is possible. Interruption of an α_2 service occurs at the completion of a concurrent β_2 service. After some time, the α_2 service continues from the point in the service at which interruption occurred. Thus, the "β_2-complete" interruption of an α_2 service is of the preemptive-resume type.

At the completion of an α_1, α_2, α_3, β_1, or β_3 service and at an interruption of an α_2 service (i.e., at completion of β_2 service), the processor chooses the next service to be provided according to a rule of priority as follows (see Figure 2.3):

(i) if there is a job waiting for β_3 service, begin this service;

(ii) otherwise, if there is a job waiting for β_1 service, begin this service provided that β_2 service is not in progress;

(iii) if the last α service provided was a completed α_2 service, begin an α_3 service;

(iv) if the last α service provided was an interrupted α_2 service, resume the α_2 service;

(v) if the last α service provided was an α_1 service, begin an α_2 service;

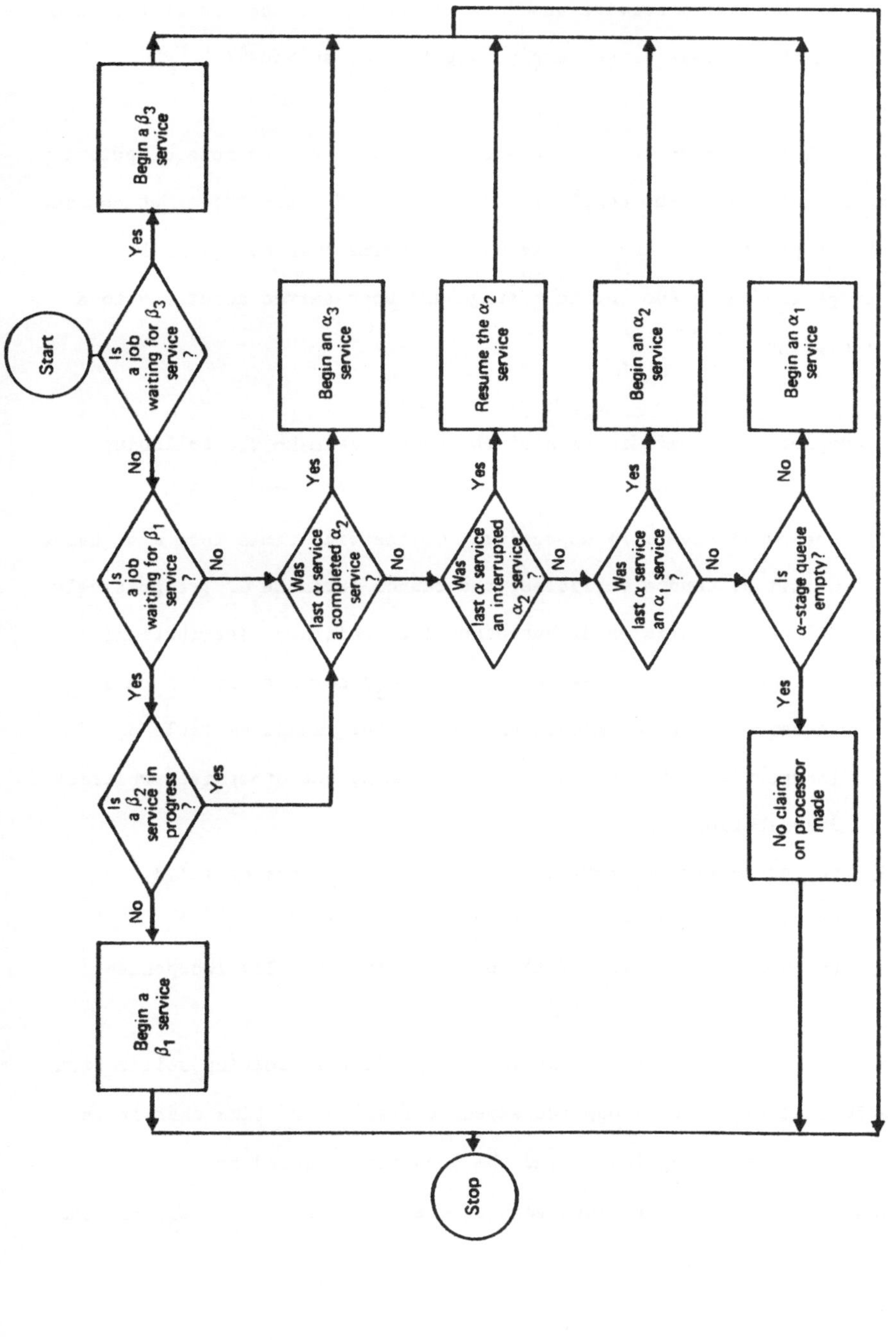

Figure 2.3. Rule of priority

(vi) if the last α service provided was an α_3 service and if the queue in the α stage is not empty, begin an α_1 service.

If no claim is made on the processor according to the rule of priority, it remains idle until the completion of the next β_2 service, at which time the rule of priority is invoked again. We assume that the queue in the α stage and the queue in the β stage are both served according to a FCFS discipline.

To complete the specification of the model, we make the following probabilistic assumptions:

(i) the successive α_1 [respectively α_3] service times form a sequence of i.i.d. random variables; the random variable α_1 [respectively α_3] has a finite mean, but otherwise arbitrary distribution;

(ii) the successive β_1 [respectively β_2, β_3] service times form a sequence of i.i.d. random variables; the random variable β_1 [respectively β_2, β_3] has a finite mean, but otherwise arbitrary distribution;

(iii) the successive α_2 service times form a sequence of i.i.d. exponentially distributed random variables;

(iv) the sequences in (i), (ii) and (iii) are mutually independent.

Quantities of interest in this model include the limiting utilization of the IO unit (i.e., the long-run expected fraction of time that it is busy as opposed to being idle), and the long-run expected fractions of time that the processor provides each of the services α_1, α_2, α_3, β_1, and

β_3. Formally, denote the total amount of α_1 [respectively α_2, α_3, β_1, β_2, β_3] service that has occurred in the system during the time interval (0,t] by $A_1(t)$ [respectively $A_2(t)$, $A_3(t)$, $B_1(t)$, $B_2(t)$, $B_3(t)$]. Then, in terms of these processes of cumulated service times, for i=1, 2, and 3, we wish to estimate

$$\lim_{t\to\infty} \frac{E\{A_i(t)\}}{t}$$

and

$$\lim_{t\to\infty} \frac{E\{B_i(t)\}}{t}, \tag{2.2.1}$$

provided that the limits exist.

The definition of an appropriate system state for a simulation of this model is perhaps somewhat less apparent than in the previous example. It is clear, however, that it is necessary to take into account more than just the number of jobs waiting or in service at one of the stages. The additional information that we use is the kind of service which is being provided by the processor or (according to the rule of priority) is to be provided next. If at time t the processor is providing α_i service ($1\le i\le 3$) and there are n jobs ($0\le n<N$) waiting or in service in the β stage, we define the system state X(t) to be (n,i). Otherwise, at time t the processor is idle or is providing β_1 or β_3 service; in this case we define X(t) to be

(i) (n,0) if there are n jobs ($0\le n\le N$) waiting or in service in the β stage and the next α service to be provided is the resumption of an α_2 service;

(ii) (n,1) if the next α service is an α_1 service; and

(iii) (n,2) if the next α service is the start of an α_2 service.

The force of this state definition is that the stochastic process $\underset{\sim}{X}=\{X(t):t\geq 0\}$ is a regenerative process in continuous time. To see this, consider the increasing sequence of time points $\{t_k:k\geq 0\}$ at which either:

(i) a β_3 service has just been completed and the served job has moved to the α stage queue, or

(ii) after a time point described in (i) at which the β stage queue is empty, the next job appears at the β stage for service.

Now consider the subsequence $\{t_k'\}$ of the $\{t_k\}$ at which the system enters state (1,1). At such a time point, there is one job in the β stage and the next α service to be provided is an α_1 service. These t_k' are regeneration points for the process $\underset{\sim}{X}$. Note that the process $\{X(t_k):k\geq 0\}$, i.e., the process $\underset{\sim}{X}$ observed at the epochs $\{t_k\}$, is an irreducible, aperiodic finite state discrete time Markov chain. In addition, it is easy to check from the basic assumptions of the model that, given the state of the system at time points t_k and t_{k+1}, the time interval $t_{k+1}-t_k$ is a random variable whose distribution does not depend on the times between previous time points t_ℓ $(\ell\leq k)$ or the state of the system at the time points t_ℓ $(\ell<k)$).

The regenerative structure of the process $\underset{\sim}{X}$ provides a basis for estimation of the quantities defined in Equation (2.2.1) even though they cannot be expressed in the form $E\{f(X)\}$. The first observation is that the sequences

$$\{(A_i(t'_{k+1})-A_i(t'_k),\ t'_{k+1}-t'_k):k\geq 1\}$$

and

$$\{(B_i(t'_{k+1})-B_i(t'_k),\ t'_{k+1}-t'_k):k\geq 1\} \tag{2.2.2}$$

consist of i.i.d. random vectors. In addition, the ratio formulas

$$\lim_{t\to\infty} \frac{E\{A_i(t)\}}{t} = \frac{E\{A_i(t'_{k+1})-A_i(t'_k)\}}{E\{t'_{k+1}-t'_k\}}$$

and

$$\lim_{t\to\infty} \frac{E\{B_i(t)\}}{t} = \frac{E\{B_i(t'_{k+1})-B_i(t'_k)\}}{E\{t'_{k+1}-t'_k\}} \tag{2.2.3}$$

hold, provided that all relevant expectations are finite. The ratio formulas can be established, for example, by direct application of a basic limit theorem for cumulative processes (Smith (1958), p. 263); the processes $\{A_i(t):t\geq 0\}$ and $\{B_i(t):t\geq 0\}$ are cumulative processes with respect to the regenerative process $\underset{\sim}{X}$. Given Equations (2.2.2) and (2.2.3), the arguments for the standard regenerative method apply and yield point estimates and confidence intervals for the quantities of interest.

This somewhat complex model illustrates the following remark about network of queues used as models for computer and communication systems. For quantities associated with queue lengths, it is often possible to define an appropriate state vector and in a fairly straightforward manner, to establish the existence of regeneration points and the applicability of the regenerative method. Frequently this is possible under fairly general distributional assumptions. The key to showing the applicability of the regenerative method, as in the example above, is often the discovery of a familiar stochastic process (e.g., discrete time Markov chain) embedded in the state vector process.

There are, however, examples of networks of queues for which, even under the convenient assumptions of distributions with a Cox-phase (exponential stage) representation for service and interarrival times, it is technically quite difficult to establish the applicability of the regenerative method; this is so even though we would expect the required conditions to be satisfied. The model of an automated tape library proposed by Lavenberg and Slutz (1975) provides one such example.

3.0. CLOSED NETWORKS OF QUEUES

We deal with closed networks of queues having a finite number of jobs (customers), N, a finite number of service centers, s, and a finite number of (mutually exclusive) job classes, c. At every epoch of (continuous) time each job is in exactly one job class, but jobs may change class as they traverse the network. Upon completion of service at center i a job of class j goes to center k and changes to class ℓ with probability $p_{ij,k\ell}$, where $\underline{P}=\{p_{ij,k\ell}:1\le i,k\le s,\ 1\le j,\ell\le c\}$ is a given irreducible Markov matrix. At each service center jobs queue and receive service according to a fixed priority scheme among classes; the priority scheme may differ from center to center. Within a class at a center, jobs receive service according to a fixed queue service discipline, e.g., first-come, first-served (FCFS). Note that in accordance with the matrix $\underline{P}$, some centers may never see jobs of certain classes. Only one job can receive service at a center at a time; i.e., the centers are single servers. According to a fixed procedure for each center, a job in service may or may not be preempted if another job of higher priority joins the queue at the center.

3.1. Probabilistic Assumptions

The marked job method to be discussed in Section 4 for passage time simulation applies to networks of this kind in which all service times in the network are mutually independent, and at a center have a distribution with a <u>Cox-phase representation</u> (Cox (1955)), i.e., consisting of a sequence of exponential stages; see Figure 3.1. We permit parameters of the service time distribution to depend on the service center, the class of job being served, and the "state" of the entire network as defined below.

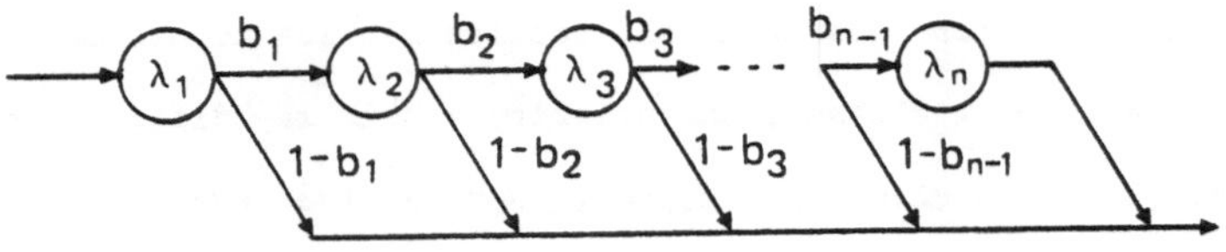

Figure 3.1. Cox-phase representation

Each service time distribution has its own finite number of stages, say n. Realization of a service time is as a sum of a random number (≥1) of exponentially distributed times. Within the jth stage ($1 \le j \le n-1$), an amount of service exponentially distributed with rate parameter λ_j accrues; with probability $1-b_j$ service is complete, and with probability b_j additional service exponentially distributed with rate parameter λ_{j+1} accrues. The density function of the resulting service time has rational Laplace transform

$$f^*(s) = \sum_{j=1}^{n} (1-b_j)a_j \prod_{k=1}^{j} \left(\frac{\lambda_k}{s+\lambda_k}\right)$$

where $a_1=1$ and for $1<j\le n$, $a_j=b_1\ldots b_{j-1}$. The class of density functions having Laplace transform of this form includes hyperexponential and mixtures of Erlang densities; see the Appendix of Gelenbe and Muntz (1976) for a discussion. Note that we exclude the case of zero service times occurring with positive probability.

In connection with this class of service time distributions, it is clear that if $b_1=b_2=\ldots=b_{n-1}=1$ and if $\lambda_1=\lambda_2=\ldots=\lambda_n$, then the resulting service time has an Erlang-n distribution, i.e., a gamma distribution with integral shape parameter. It is less obvious but can be shown (by considering Laplace transforms of the density functions) that if $\lambda_1>\lambda_2$ and

$$b_1 = 1 - p_1 - p_2\lambda_2/\lambda_1 ,$$

then the Cox-phase form is a representation for the hyperexponential density

$$p_1\lambda_1 e^{-\lambda_1 t} + p_2\lambda_2 e^{-\lambda_2 t} ,$$

concentrated on $t \geq 0$, where p_1 and p_2 are nonnegative and $p_1+p_2=1$. The corresponding result for three stages is that if $\lambda_1>\lambda_2>\lambda_3$,

$$b_1 = 1 - p_1 - (p_2\lambda_2/\lambda_1) - (p_3\lambda_3/\lambda_1) ,$$

and

$$b_2 = 1 - [p_2+\{p_3\lambda_3(\lambda_1-\lambda_3)\}/\{\lambda_2(\lambda_1-\lambda_2\}]/[p_2+\{p_3(\lambda_1-\lambda_3)/(\lambda_1-\lambda_2)\}] ,$$

then the Cox-phase form is a representation for the hyperexponential density

$$p_1\lambda_1 e^{-\lambda_1 t} + p_2\lambda_2 e^{-\lambda_2 t} + p_3\lambda_3 e^{-\lambda_3 t} ,$$

where p_1, p_2, and p_3 are nonnegative and $p_1+p_2+p_3=1$.

3.2. State Vector Definition

In order to characterize the state of the network at time t, we let $S_i(t)$ denote the class of the job receiving service at center i at time t, where $i=1,2,\ldots,s$; by convention $S_i(t)=0$ if at time t there are no jobs at center i. The classes of jobs serviced at center i ordered by decreasing priority are $j_1(i),j_2(i),\ldots,j_{k(i)}(i)$, elements of the set $\{1,2,\ldots,c\}$. Let $C_{j_1}^{(i)}(t),\ldots,C_{j_{k(i)}}^{(i)}(t)$ denote the number of jobs in queue at time t of the various classes of jobs serviced at center i, $i=1,2,\ldots,s$. For queue lengths of jobs of various classes at the several centers, these state variables (together with the stages-of-service) would suffice. To deal with general characteristics of passage times, however, these state variables are inadequate. An apparently minimal state vector augmentation is based on the concept of a marked job. The idea is to keep track of the position of an arbitrarily chosen marked job in the network and to measure its passage times during the simulation. It is convenient to think of the N jobs as being completely ordered in a linear stack according to the following scheme. For $t \geq 0$, we define the vector $Z(t)$ by

$$Z(t) = (C^{(1)}_{j_{k(1)}}(t), C^{(1)}_{j_{k(1)-1}}(t), \ldots, C^{(1)}_{j_1}(t), S_1(t); \ldots;$$

$$C^{(s)}_{j_{k(s)}}(t), C^{(s)}_{j_{k(s)-1}}(t), \ldots, C^{(s)}_{j_1}(t), S_s(t)) \, . \qquad (3.2.1)$$

The linear job stack then corresponds to the order of components in the vector $Z(t)$ after ignoring any zero components. Within a class at a particular service center, jobs waiting appear in the job stack in FCFS order, i.e., in order of their arrival at the center, the latest to arrive being closest to the top of the stack. We denote by $N(t)$ the position (from the top) of the marked job in this job stack at time t.

We associate a stage of service with each job in the network as follows. A job in service is in a particular stage of its service time distribution at that center; for such a job, this is the associated stage. For a job in queue, the associated stage is the stage of its service time distribution that is to be provided when the job next receives service; typically this is the first stage of service, but may be a subsequent stage if the job has been preempted. For $t \geq 0$, we define the vector $U(t)$ by

$$U(t) = (U_1(t), \ldots, U_N(t)) \, , \qquad (3.2.2)$$

where $U_j(t)$ is the stage of service associated with the jth job in the linear stack of jobs, and for $t \geq 0$ take as the state vector of the network of queues the vector

$$X(t) = (Z(t),N(t),U(t)) \ . \qquad (3.2.3)$$

For any service center i that sees only one class of job, i.e., such that $k(i)=1$, it is possible to simplify the state vector by replacing $C^{(i)}_{j_{k(i)}}(t),S_i(t)$ by $Q_i(t)$, the total number of jobs at center i. Note that the state vector definition does not explicitly take into account that the total number of jobs in the network is fixed. In the case of complex networks, the use of this resulting somewhat larger state space facilitates generation of the state vector process; for relatively simple networks, it may be desirable to remove the redundancy.

3.3. Definition of Passage Times

Given a particular closed network of queues, we must specify the passage time of interest. This can be done in terms of the arbitrarily chosen marked job of Section 3.2, by means of four subsets (A_1, A_2, B_1, B_2) of the state space, E, of the stochastic process $\underset{\sim}{X}=\{X(t):t\geq 0\}$. The sets A_1, A_2 [respectively B_1, B_2] jointly define the random times at which passage times for the marked job start [respectively terminate]. The sets A_1, A_2, B_1 and B_2 in effect determine when to start and stop the clock measuring a particular passage time of the marked job.

It is convenient to introduce the jump times $\{\tau_n:n\geq 0\}$ of the process $\underset{\sim}{X}$. We set $\tau_0=0$ and have $\tau_0<\tau_1<\ldots$ with probability one. Since the state space E of $\underset{\sim}{X}$ is finite, there can be no finite accumulation points for the τ_n and $\tau_n \to \infty$. For $k, \ell_1, n \geq 1$, we require that the sets A_1, A_2, B_1 and B_2 satisfy:

if $X(\tau_{n-\ell_1}) \in A_1$, $X(\tau_n) \in A_2$, $X(\tau_{n-\ell_2+k}) \in A_1$ and $X(\tau_{n+k}) \in A_2$,

then $X(\tau_{n-\ell_3+m}) \in B_1$ and $X(\tau_{n+m}) \in B_2$ for some $0 < m \leq k$ and $1 \leq \ell_3 \leq m$;

and

if $X(\tau_{n-\ell_1}) \in B_1$, $X(\tau_n) \in B_2$, $X(\tau_{n-\ell_2+k}) \in B_1$ and $X(\tau_{n+k}) \in B_2$,

then $X(\tau_{n-\ell_3+m}) \in A_1$ and $X(\tau_{n+m}) \in A_2$ for some $0 \leq m < k$ and $1 \leq \ell_3 \leq m+\ell_1$. (3.3.1)

These conditions ensure that the start and termination times for the specified passage time strictly alternate.

In terms of the sets A_1, A_2, B_1, and B_2, we define two sequences of random times, $\{S_j : j \geq 0\}$ and $\{T_j : j \geq 1\}$, where S_{j-1} is the start time of the jth passage time for the marked job and T_j is the termination time of this jth passage time. Assuming that the initial state of the process $\underset{\sim}{X}$ is such that a passage time for the marked job starts at t=0, let

$$S_0 = 0$$

$$S_j = \inf\{\tau_n \geq T_j : X(\tau_n) \in A_2,\ X(\tau_k) \in A_1 \text{ for some } \tau_k \geq S_{j-1} \text{ and } k < n\},\ j \geq 1 ,$$

and

$$T_j = \inf\{\tau_n > S_{j-1} : X(\tau_n) \in B_2,\ X(\tau_k) \in B_1 \text{ for some } \tau_k \geq S_{j-1} \text{ and } k < n\},\ j \geq 1 . \quad (3.3.2)$$

Then the jth passage time for the marked job is simply $P_j=T_j-S_{j-1}$, $j\geq 1$. Quite often (and in particular in the two examples discussed in Section 5), the value of k in the definitions of S_j and T_j can be set equal to n-1, a considerable simplification. In general, however, we allow $k<n$. For response times (complete circuits), $A_1=B_1$ and $A_2=B_2$; consequently $S_j=T_j$ for all $j\geq 1$.

The following example illustrates specification of passage times in terms of the sets A_1, A_2, B_1, and B_2.

(3.3.3) EXAMPLE. Consider the closed network of Figure 3.2 having three service centers. Upon completion of α [respectively β] service at center 1 [respectively center 2], each of the N jobs joins the tail of the queue in center 3. In accordance with a binary-valued variable ψ, after completion of γ service at center 3, a job joins the tail of the queue in center 1 (when $\psi=1$) or joins the tail of the queue in center 2 (when $\psi=0$). Assume that each of the queues is served according to a FCFS discipline. Also assume that routing variable is a random variable independent of service times in the network, and that at each of the centers, service times are i.i.d. exponential random variables.

Suppose that the passage time of interest starts when a job joins the center 1 queue (upon completion of service at center 3) and terminates when the job joins the queue in center 3. Here there is effectively a single job class. Denoting by p the (independent) probability that the binary routing variable ψ takes the value 1, in the routing matrix $\underline{P}$, $p_{11,31}=p_{21,31}=1$,

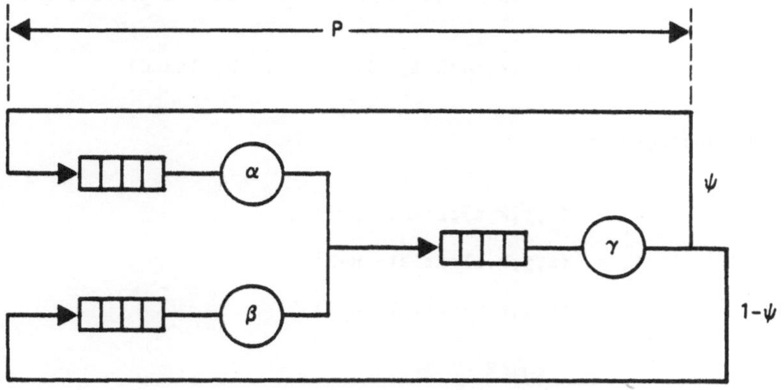

Figure 3.2. Closed network of queues with three service centers

$p_{31,11}=p$, $p_{31,21}=1-p$, and all other $p_{ij,k\ell}=0$. For $t\geq 0$ and $i=1,2$, and 3, we define $Q_i(t)$ to be the number of jobs waiting or in service at center i. Then

$$X(t) = (Z(t),N(t)) \ ,$$

where $Z(t)=(Q_1(t),Q_2(t),Q_3(t))$ and $N(t)$ is the position of the marked job in the associated job stack. The state space E of the process $\underset{\sim}{X}$ is

$$E = \{(i,j,k,n):0\leq i,j,k\leq N;\ i+j+k=N;\ 1\leq n\leq N\} \ .$$

For this passage time

$$A_1 = \{(i,j,k,n)\in E:k\geq 1;\ n=N\} \ ,$$

$$A_2 = \{(i,j,k,n)\in E:i\geq 1;\ n=1\} \ ,$$

$$B_1 = \{(i,j,k,n)\in E:i\geq 1;\ n=i\} \ ,$$

and

$$B_2 = \{(i,j,k,n)\in E:k\geq 1;\ n=i+j+1\} \ .$$

Now assume that the queues in center 1 and center 2 are served according to a FCFS discipline, but that at center 3, jobs that join the queue after completion of service at center 1 have priority over those that join the queue after completion of service at center 2. Suppose that the passage time (indicated by P in Figure 3.2) of interest starts when a job joins the queue in center 1 and terminates at the completion of service to the job at center 3.

To deal with this passage time, we use two job classes: center 1 sees only jobs of class 1, center 2 sees only jobs of class 2, and center 3 sees jobs of class 1 and class 2. In the routing matrix $\underline{P}$, $p_{11,31}=p_{22,32}=1$, $p_{31,11}=p_{32,11}=p$, $p_{31,22}=p_{32,22}=1-p$, and all other $p_{ij,k\ell}=0$. For $t\geq 0$,

$$Z(t) = (Q_1(t),Q_2(t),C_2^{(3)}(t),C_1^{(3)}(t),S_3(t)) ,$$

where $C_1^{(3)}(t)$ [respectively $C_2^{(3)}(t)$] is the number of class 1 [respectively class 2] jobs in queue at center 3 at time t, and $S_3(t)$ is the class of job in service at center 3 at time t. ($S_3(t)=0$ if at time t there are no jobs at center 3). The state space E of the process $\underset{\sim}{X}=\{X(t):t\geq 0\}$ is

$$E = \{(i,j,0,0,0,n):0\leq i,j\leq N;\ i+j=N;\ 1\leq n\leq N\} \cup \{(i,j,k,\ell,m,n):0\leq i,j,k,\ell\leq N;\ i+j+k+\ell=N-1;\ 1\leq m\leq 2;\ 1\leq n\leq N\}$$

For the passage time P,

$$A_1 = \{(i,j,k,\ell,m,n)\in E:m=1;\ n=N\} ,$$

$$A_2 = \{(i,j,k,\ell,m,n)\in E:i\geq 1;\ n=1\} ,$$

$$B_1 = A_1 ,$$

and

$$B_2 = A_2 \cup \{(i,j,k,\ell,m,n)\in E:j\geq 1;\ n=i+1\} .$$

4.0. THE MARKED JOB METHOD

We base the estimation of general characteristics of passage times in a closed network of queues of Section 3 on the measurement of passage times for a typical job, the marked job discussed above. It is intuitively clear and is shown in Appendix 1 that the sequence of passage times for any other job (as well as the sequence of passage times, irrespective of job identity, in order of start or termination) converges in distribution to the same random variable as the sequence of passage times for the marked job. It follows that we can estimate general characteristics of passage times in the network by simulation of the process $\underset{\sim}{X}=\{X(t):t\geq 0\}$ defined by Equations (3.2.1), (3.2.2) and (3.2.3). We shall see that it is possible to use the regenerative method (applied to an appropriate discrete time regenerative process) to obtain strongly consistent point estimates and asymptotically valid confidence intervals for passage time characteristics.

We denote by X_n, $n\geq 0$, the state of the system when the (n+1)st passage time of the marked job starts. For $j\geq 1$, let P_j be the jth passage time for the marked job and take the quantity of interest in the simulation to be

$$r(f) = E\{f(X,P)\} , \tag{4.0.1}$$

where f is a real-valued (measurable) function and $(X_n,P_{n+1})\Rightarrow(X,P)$. For example, to estimate $E\{P\}$, we take $f(i,x)=x$; to estimate $P\{P\leq t\}$, we take $f(i,x)=1_{[0,t]}(x)$, where $1_{[0,t]}$ is the indicator function of the set $[0,t]$.

4.1. Simulation for Passage Times

Based on the measurement of passage times for the marked job, we obtain estimates of the quantity r(f) as follows.

(4.1.1) ALGORITHM. (Marked Job Method)

1. To serve as a return state, select a state of the system, i_0, at which a passage time of the marked job starts. Begin the simulation with $X(0)=i_0$.
2. Carry out the simulation of $\underset{\sim}{X}$ for a fixed number, n, of cycles (having random length) defined by the successive returns to the state i_0.
3. In each cycle measure all passage times of the marked job. Denote the number of passage times of the marked job observed in the kth cycle by $\alpha_k (k\geq1)$, and compute the quantity

$$Y_k(f) = \sum_{n=\beta_{k-1}}^{\beta_k-1} f(X_n,P_{n+1}) , \tag{4.1.2}$$

where $\beta_0=0$ and $\beta_k=\alpha_1+...+\alpha_k$.

4. Take as a point estimate (based on n cycles) of r(f) defined by Equation (4.0.1) the quantity

$$\hat{r}_n(f) = \overline{Y}_n(f)/\overline{\alpha}_n , \tag{4.1.3}$$

where

$$\overline{Y}_n(f) = n^{-1} \sum_{k=1}^{n} Y_k(f)$$

and

$$\overline{\alpha}_n = n^{-1} \sum_{k=1}^{n} \alpha_k .$$

5. Take as a $100(1-2\gamma)\%$ confidence interval (based on n cycles) for r(f) the interval

$$\hat{I}_n(f) = \left[\hat{r}_n(f)-z_{1-\gamma}s_n/\left(\bar{\alpha}_n n^{1/2}\right),\ \hat{r}_n(f)+z_{1-\gamma}s_n/\left(\bar{\alpha}_n n^{1/2}\right)\right] \quad . \tag{4.1.4}$$

Here $z_{1-\gamma}=\Phi^{-1}(1-\gamma)$, where $\Phi(\cdot)$ is the distribution function of a standardized (mean zero, variance one) normal random variable, and s_n is the quantity

$$s_n = \left\{s_{11}-2\hat{r}_n(f)s_{12}+\left(\hat{r}_n(f)\right)^2 s_{22}\right\}^{1/2}$$

where

$$s_{11} = (n-1)^{-1}\sum_{k=1}^{n}(Y_k(f)-\bar{Y}_n(f))^2 ,$$

$$s_{22} = (n-1)^{-1}\sum_{k=1}^{n}(\alpha_k-\bar{\alpha}_n)^2$$

and

$$s_{12} = (n-1)^{-1}\sum_{k=1}^{n}(Y_k(f)-\bar{Y}_n(f))(\alpha_k-\bar{\alpha}_n) .$$

Justification for this passage time simulation method appears in the next section.

4.2. The Underlying Stochastic Structure

With knowledge of the matrix $\underline{P}$ and the parameters of the Cox-phase service times, we can carry out the simulation (i.e., generation of the process $\underset{\sim}{X}=\{X(t):t\geq 0\}$ as defined in Section 3.2) in a straightforward manner. Note that when all service times are exponentially distributed, the vector U(t) in the state vector given by Equation (3.2.3) can be omitted, resulting in a smaller, less complex state space.

The force of the assumptions of Cox-phase service times and Markovian routing in the closed networks of queues of Section 3.1 is contained in Theorem (4.2.1).

(4.2.1) THEOREM. The stochastic process $\underset{\sim}{X}=\{X(t):t\geq 0\}$ defined by Equations (3.2.1), (3.2.2) and (3.2.3) is a continuous time Markov chain with a finite state space, E.

The method of Section 4.1 for estimation of general characteristics of passage times relies on the measurement of passage times for a typical job, the marked job as introduced in Section 3.2. As in Section 3.3, specification of the passage time of interest is in terms of the marked job, by means of the four subsets (A_1, A_2, B_1, B_2) of E. We define the two sequences of random times, $\{S_j:j\geq 0\}$ and $\{T_j:j\geq 1\}$, where S_{j-1} is the start time of the jth passage time for the marked job and T_j is the termination time of this jth passage time. For an initial state of the Markov chain $\underset{\sim}{X}$ such that a passage time for the marked job begins at t=0,

$$\begin{aligned} S_0 &= 0 \\ S_j &= \inf\{\tau_n \geq T_j : X(\tau_n)\in A_2,\ X(\tau_k)\in A_1 \text{ for some } \\ &\qquad \tau_k \geq S_{j-1} \text{ and } k<n\},\ j\geq 1 \end{aligned}$$

and

$$\begin{aligned} T_j &= \inf\{\tau_n > S_{j-1} : X(\tau_n)\in B_2,\ X(\tau_k)\in B_1 \text{ for some } \\ &\qquad \tau_k \geq S_{j-1} \text{ and } k<n\},\ j\geq 1\ , \end{aligned} \tag{4.2.2}$$

where $\{\tau_n:n\geq 0\}$ are the jump times of the Markov chain $\underset{\sim}{X}$. Then the jth passage time for the marked job is $P_j=T_j-S_{j-1}$, $j\geq 1$.

Let X_n denote the state of the continuous time Markov chain $\underset{\sim}{X}=\{X(t):t\geq 0\}$ when the (n+1)st passage time of the marked job starts: $X_n=X(S_n)$, $n\geq 0$. Since $\underset{\sim}{X}$ is a Markov chain and $\{S_n:n\geq 0\}$ are stopping times for the chain, $\{X_n:n\geq 0\}$ is a discrete time Markov chain with finite state space A_2. For a discussion of stopping times, see Çinlar (1975a), p. 239. Throughout, we shall assume that the process $\{X_n:n\geq 0\}$ is irreducible and aperiodic. Furthermore, it is easy to check that the process $\{(X_n,S_n):n\geq 0\}$ satisfies

$$P\{X_{n+1}=j,\ S_{n+1}-S_n\leq t|X_0,\ldots,X_n;\ S_0,\ldots,S_n\}$$
$$=P\{X_{n+1}=j,\ S_{n+1}-S_n\leq t|X_n\}$$

with probability one for all $n\geq 0$, $j\epsilon A_2$, and $t\geq 0$.

(4.2.3) THEOREM. The stochastic process $\{(X_n,S_n):n\geq 0\}$ is a Markov renewal process.

This follows directly from the definition of a Markov renewal process; see Çinlar (1975b) or Çinlar (1975a), p. 313. The basic data for this Markov renewal process is the semi-Markov kernel over A_2, $\underset{\sim}{K}=\{K(i,j;t):i,j\epsilon A_2,\ t\geq 0\}$, defined by

$$K(i,j;t)=P\{X_{n+1}=j,\ S_{n+1}-S_n\leq t|X_n=i\}\ .$$

While the kernel $\underset{\sim}{K}$ is normally given in the analysis of a Markov renewal process, for the network of queues passage time problem, $\underset{\sim}{K}$ is virtually impossible to calculate. Thus from this point of view, the only hope for studying the passage time in question is to generate sample paths of $\{(X_n,S_n):n\geq 0\}$ via simulation of the Markov chain $\underset{\sim}{X}$.

As in Section 4.0, let f be a real-valued (measurable) function with domain $A_2\times R_+$, where $R_+=[0,\infty)$, and define the quantity r(f) according to

$$r(f) = E\{f(X,P)\} . \tag{4.2.4}$$

Now we select a fixed state, i_0, from A_2, begin the simulation of $\underset{\sim}{X}$ with $X(0)=i_0$, and carry out the simulation of $\underset{\sim}{X}$ in cycles defined by the successive returns to state i_0. Let α_k denote the length (in discrete time units) of the kth cycle of $\{X_n:n\geq 0\}$ and define $\beta_0=0$, and $\beta_k=\alpha_1+\ldots+\alpha_k$, $k\geq 1$.

Theorem (4.2.5) follows from Theorem (4.2.3) and the definition of a regenerative process. It comprises the key observation which leads to point and interval estimates for r(f).

(4.2.5) THEOREM. The stochastic process $\{(X_n,P_{n+1}):n\geq 0\}$ is a regenerative process in discrete time with regeneration points $\{\beta_k:k\geq 0\}$.

Note that the regeneration points β_k are not the times of return to a fixed state of the process $\{(X_n,P_{n+1}):n\geq 0\}$.

For $k\geq 1$, we now define

$$Y_k(f) = \sum_{n=\beta_{k-1}}^{\beta_k-1} f(X_n,P_{n+1}) . \tag{4.2.6}$$

Since the β_k are regeneration points for $\{(X_n,P_{n+1}):n\geq 0\}$, we have (cf. Crane and Iglehart (1975), Proposition 2.1) the following result.

(4.2.7) THEOREM. The pairs of random variables $\{(Y_k(f),\alpha_k):k\geq 1\}$ are independent and identically distributed.

The regenerative property guarantees (Miller (1972)) that as $n\to\infty$

$$(X_n,P_{n+1}) \Rightarrow (X,P) ,$$

i.e., there exist random variables X and P such that

$$\lim_{n\to\infty} P\{X_n=i,\ P_{n+1}\le x\} = P\{X=i,\ P\le x\} \equiv g(i,x)$$

for all $i\in A_2$ and $x\in R_+$ for which $g(i,\cdot)$ is continuous. For the function f (appearing in Equation (4.2.4)) in the definition of r(f), let D(f) denote the set of (i,x) pairs at which $f(i,\cdot)$ is discontinuous. Assuming that $P\{(X,P)\in D(f)\}=0$, it follows that

$$f(X_n,P_{n+1}) \Rightarrow f(X,P) \qquad (4.2.8)$$

as $n\to\infty$. The final step is to establish a ratio formula for $E\{f(X,P)\}$ which makes it possible to apply the regenerative method to estimation of passage times; this follows from the general result for regenerative processes (cf. Crane and Iglehart (1975), Proposition A.3). A direct proof not requiring the key renewal theorem is in Appendix 2.

(4.2.9) THEOREM. Assume that $E\{|f(X,P)|\}<\infty$. Then

$$E\{f(X,P)\}=E\{Y_1(f)\}/E\{\alpha_1\} ,$$

where $Y_1(f)$ is given by Equation (4.2.6).

With the ratio formula of Theorem (4.2.9) and the fact that the pairs of random variables $\{(Y_k(f),\alpha_k):k\ge 1\}$ are independent and identically distributed (Theorem (4.2.7)), we can apply the regenerative method to $\{(X_n,P_{n+1}):n\ge 0\}$ to obtain point and interval estimates for r(f).

5.0. EXAMPLES AND SIMULATION RESULTS

To illustrate the marked job method of the previous sections for estimation of passage times in closed networks of queues, we consider two examples. Descriptions of the networks and state vector definitions appear in this section, along with illustrative numerical results.

5.1. A Closed Network of Queues

The first example is relatively simple; see Figure 5.1. Upon completion of α service at center 1, in accordance with a binary-valued variable ψ, the job joins the tail of the queue in center 1 (when $\psi=1$) or joins the tail of the queue in center 2 for β service (when $\psi=0$). After completion of β service at center 2, the job joins the tail of the queue in center 1. Neither center 1 nor center 2 service is subject to interruption. We assume that both queues are served according to an FCFS discipline. The limiting response time of interest (denoted by R) for a job is the time measured from entrance into the center 1 queue upon completion of a center 2 service until the job next joins the center 1 queue. Also of possible interest in this model is the limiting passage time (denoted by P) corresponding to the time measured from entrance into the center 1 queue upon completion of a center 2 service until the job next joins the center 2 queue.

We consider passage time simulation of this network of queues under the following <u>probabilistic assumptions</u>:

(i) successive α service times form a sequence of i.i.d. random variables, exponentially distributed with rate parameter μ_0;

(ii) successive β service times form a sequence of i.i.d. random variables, exponentially distributed with rate parameter μ_1;

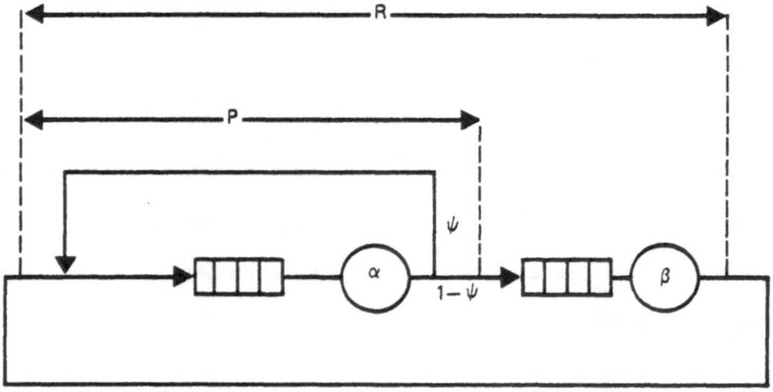

(i) α and β services are not interruptable
(ii) Routing determined by binary valued random variable ψ

Figure 5.1. Closed network of queues

(iii) ψ is a Bernoulli random variable and values of ψ at successive α service completions form a sequence of i.i.d. random variables;

(iv) the sequences in (i), (ii) and (iii) are mutually independent.

In this network, there are two classes of jobs: class 1 jobs at center 1 and class 2 jobs at center 2. Since each center sees only one class of job, the process $\{Z(t):t\geq 0\}$ can be defined by

$$Z(t) = (Q_1(t),Q_2(t))$$

where

$Q_1(t)$ = number of jobs waiting or in service at center 1 at time t,

and

$Q_2(t)$ = number of jobs waiting or in service at center 2 at time t.

Now let N(t) denote the position of the marked job in the linear job stack corresponding to the order of the nonzero components of Z(t). Taking into account the fixed number of jobs in the network and that the service times are exponentially distributed, the process $\underset{\sim}{X}=\{X(t):t\geq 0\}$, where

$$X(t) = (Q_1(t),N(t)) ,$$

is the underlying continuous time Markov chain.

For this model the state space E of $\underset{\sim}{X}$ is

$$E = \{(i,j):0\leq i\leq N;\ 1\leq j\leq N\} ,$$

where N is the number of jobs in the network. In the special case of N=2 jobs, Figure 5.2 shows the possible state transitions among the

$N\times(N+1)=6$ states. State (0,1) corresponds to both jobs in center 2 with the marked job being the one in queue; state (1,1) corresponds to one job in service at center 1 and one in service at center 2, the marked job being the former, etc.

The Markov chain governing change of job class at service completions can be taken to have state space {1,2} and transition matrix

$$\underline{P} = \begin{bmatrix} p & 1-p \\ 1 & 0 \end{bmatrix},$$

where we denote by p the probability that the binary branching variable ψ takes the value 1.

The sets A_1 and A_2 (of Section 4.2) defining the start of response times for the marked job are

$$A_1 = \{(i,j)\in E: j=N,\ i<N\}$$

and

$$A_2 = \{(i,j)\in E: j=1,\ i>0\}\ .$$

Since R is a response time, $T_j=S_j$ for $j\geq 1$, $B_1=A_1$, and $B_2=A_2$. For $N=2$ jobs, $A_1=\{(0,2),\ (1,2)\}$ and $A_2=\{(1,1),\ (2,1)\}$; see Figure 5.2. A sample path of the Markov renewal process $\{(X_n,S_n):n\geq 0\}$ is in Figure 5.3. For the passage time P, $A_1=\{(0,2),\ (1,2)\}$, $A_2=\{(1,1),\ (2,1)\}$, $B_1=\{(1,1),\ (2,2)\}$ and $B_2=\{(0,1),\ (1,2)\}$.

5.2. A Computer System Model

The second example is more complex, and is essentially the network of queues defined by Lavenberg and Shedler (1976) as a model of resource

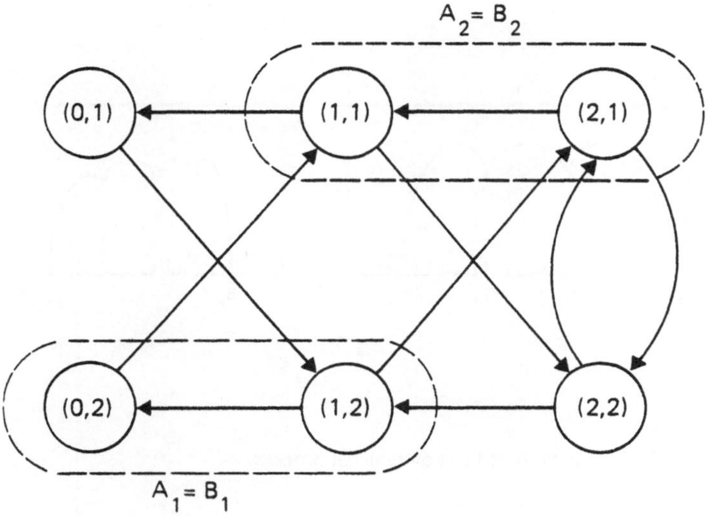

Figure 5.2. State transitions in Markov chain X and subsets of E for response time R

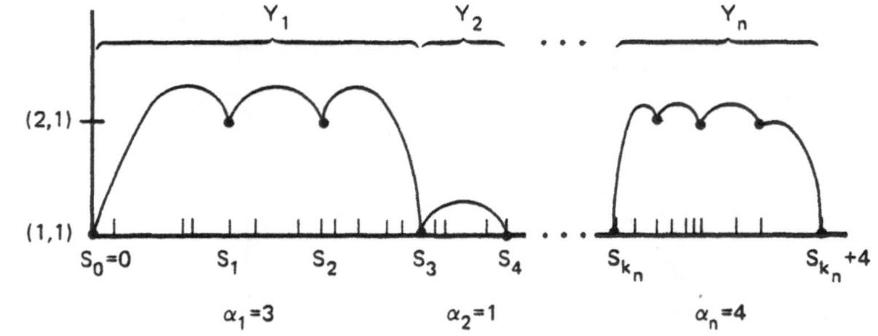

Figure 5.3. Sample Path of Markov renewal process

contention in the so-called "DL/I component" of an IMS (Information Management System) data base management computer system; see Figure 5.4. The interpretation of this figure differs from that of a conventional queueing network diagram in that services are distinguished from the servers which perform them. Thus, circles in Figure 5.4 represent services rather than servers. Five types of service, denoted by α_0, α_1, α_2, α_3 and β are represented in this model. The α services (α_0, α_1, α_2 and α_3) are performed by a single server, interpreted as the processor, and the β service is performed by a single server, interpreted as an input-output (I/O) unit. We assume in this model that no two α services can be in progress concurrently, but that any α service can be in progress concurrently with a β service. We also assume that each of the α_0, α_1, α_2 and β services is noninterruptable. The α_3 service is, however, interruptable at the completion of a β service, this interruption being of the preemptive-resume type.

A fixed number of jobs circulate in the network from time zero. At any subsequent instant of time, a job either is receiving service or waiting for service in one of the queues. We assume that each of the queues is served according to an FCFS discipline. Note that there are two queues (denoted by $q_{1,1}$ and $q_{1,2}$) for α_1 service and two queues ($q_{2,1}$ and $q_{2,2}$) for α_2 service. The arrows in Figure 5.4 indicate the flow of jobs. There are two branches leaving the α_1 service and two branches leaving the α_2 service; these branches are labeled by binary-valued variables ψ_1 and ψ_2. Upon completion of an α_1 or α_2 service, the job just

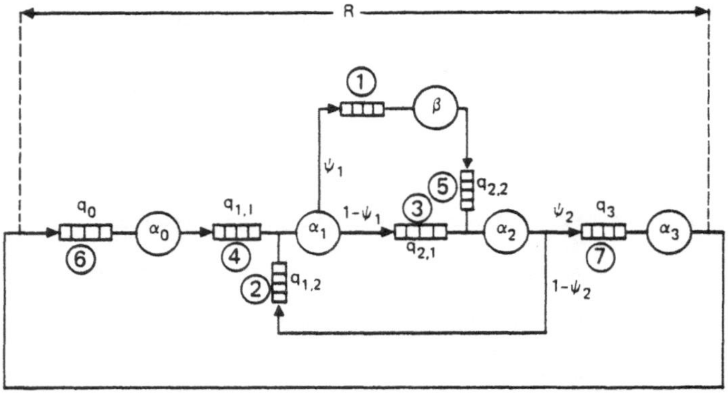

(i) Processor provides α_0, α_1, α_2 and α_3 services
(ii) I/O unit provides β service
(iii) α_0, α_1, α_2 and β services are not interruptable
(iv) α_3 service has pre-emptive resume type interruption at completion of β service
(v) Processor scheduled according to priority ordering of queues q_0, $q_{1,1}$, $q_{1,2}$, $q_{2,1}$, $q_{2,2}$ and q_3
(vi) Routing determined by binary valued random variables ψ_1 and ψ_2

Figure 5.4. DL/I component model

served follows the branch with a label having value 1. Thus, e.g., if upon completion of an α_1 service the current value of ψ_1 is zero, the job just served joins queue $q_{2,1}$.

An epoch of completion of any α service, or an epoch of completion of a β service at which either no α service is in progress or an α_3 service is in progress, is called a _scheduling decision epoch_. At such a time point, the next processor service is determined by a _processor scheduling algorithm_. Upon completion of a service, the served job immediately joins the next queue on its route. The next processor service is the service having highest priority, this priority being determined by a total ordering of queues q_0, $q_{1,1}$, $q_{1,2}$, $q_{2,1}$, $q_{2,2}$ and q_3. The processor scheduling algorithm employs the total ordering $q_{1,2}$, $q_{2,1}$, $q_{1,1}$, $q_{2,2}$, q_0, q_3 (highest to lowest priority).

The limiting passage time of interest in the DL/I component model is indicated by R in Figure 5.4. It is interpreted as the response time to a "DL/I call" in an IMS data base management system.

We consider passage time simulation of this network under the following _probabilistic assumptions_:

(i) for i=0,1,2, and 3, successive α_1 service times form a sequence of i.i.d. random variables;

(ii) successive β service times form a sequence of i.i.d. random variables;

(iii) the branching at successive α_1 (respectively α_2) service completions is governed by a sequence of i.i.d. Bernoulli random variables ψ_1 (respectively ψ_2);

(iv) the sequences in (i), (ii), and (iii) are mutually independent;

(v) the random variables α_0, α_1, α_2, α_3 and β are exponentially distributed;

(vi) $P\{\psi_2=1\}>0$, so that jobs eventually receive α_3 service.

The circled numbers in Figure 5.4 are classes of jobs in the indicated queues and are the key to representing the DL/I component model as a queueing network of the kind described in Section 3.0. This representation is shown in Figure 5.5. Classes 2-7 are identified with the α center server (processor) and class 1 with the β center server (I/O unit). At the α center server, classes 2-7 are ordered according to priority. Class 7 service is interruptable, but all other classes of service are not interruptable.

In the DL/I component model, for $t\geq 0$ we define the vector $Z(t)$ by

$$Z(t) = (Q(t),C_7(t),\ldots,C_2(t),S(t)) , \qquad (5.2.1)$$

where

$Q(t)$ = number of jobs waiting or in service at β center at time t,

$S(t)$ = class of job being served by α center server at time t,

and for $2\leq i\leq 7$,

$C_i(t)$ = number of jobs of class i waiting for service at α center at time t.

We let $N(t)$ denote the position at time t of the marked job in the job stack corresponding to the order of the nonzero components of $Z(t)$. Then

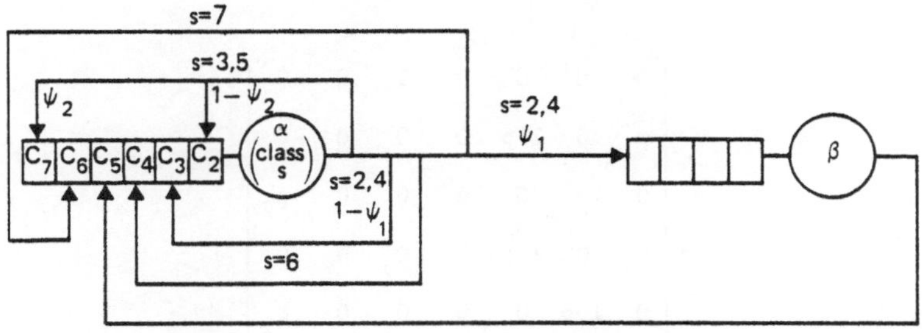

Figure 5.5. DL/I component model with job classes

the process $\underset{\sim}{X}=\{X(t):t\geq 0\}$, where

$$X(t) = (Z(t),N(t)) \ , \tag{5.2.2}$$

is the underlying continuous time Markov chain.

The Markov chain governing change of job classes at service completions can be taken to have state space $\{1,2,\ldots,7\}$ and transition matrix by

$$\underline{P} = \begin{bmatrix} 0 & 0 & 0 & 0 & 1 & 0 & 0 \\ p_1 & 0 & 1-p_1 & 0 & 0 & 0 & 0 \\ 0 & 1-p_2 & 0 & 0 & 0 & 0 & p_2 \\ p_1 & 0 & 1-p_1 & 0 & 0 & 0 & 0 \\ 0 & 1-p_2 & 0 & 0 & 0 & 0 & p_2 \\ 0 & 0 & 0 & 1 & 0 & 0 & 0 \\ 0 & 0 & 0 & 0 & 0 & 1 & 0 \end{bmatrix}$$

where p_1 [respectively p_2] denotes the probability that the binary branching variable ψ_1 [respectively ψ_2] takes the value 1.

The sets A_1 and A_2 defining the start of response times for the marked job are

$$A_1 = \{(i,N-(i+1),0,0,0,0,0,7,N):0\leq i<N\}$$

and

$$A_2 = \{(i,N-(i+1),0,0,0,0,0,6,N):0\leq i<N\} \ .$$

Since R is a response time, $T_j=S_j$ for $j\geq 1$, and $B_1=A_1$ and $B_2=A_2$.

Note that if we modify the DL/I component model so that there is no preemption of α_3 service, then the sets A_1 and A_2 are

$$A_1 = \{(i,N-(i+1),0,0,0,0,0,7,N):0\le i<N\}$$

and

$$A_2 = \{(i-m,N-(i+1),1,m-1,0,0,0,5,N-m):1\le i<N;\ 1\le m\le i\}$$
$$\cup\ \{(i,N-(i+1),0,0,0,0,0,6,N):0\le i<N\}\ .$$

This provides an example of the desirability of allowing $k<n$ in the formal definitions of the times S_j and T_j given in Section 4.2.

5.3. Numerical Results

The results displayed here were obtained using the particular linear congruential uniform random number generator described by Lewis, Goodman and Miller (1969). Exponential service times were generated by logarithmic transformation of the uniform random numbers; independent streams of exponential random numbers (resulting from different seeds of the uniform random number generator) were used to generate individual exponential holding time sequences.

The numerical values of point estimates and confidence intervals reported here for $r(f)=E\{Y_1(f)\}/E\{\alpha_1\}$ are the classical ratio estimates, i.e., use the point estimator $\hat{r}_n(f)$ of Equation (4.1.3) and the $100(1-2\gamma)\%$ confidence interval $\hat{I}_n(f)$ of Equation (4.1.4).

Results obtained for the response time R by simulation of the closed network of queues of Figure 5.1 are in Table 5.1. The initial state for the Markov chain $\underset{\sim}{X}$ (and return state identifying cycles) is the state (1,1). The results in Table 5.1 are for $N=2$ jobs, with $\mu_0=1$, $\mu_1=0.5$ and $p=0.75$. Theoretical values for the long-run expected fractions of time that centers 1 and 2 are busy (utilizations), and the expected response time can be obtained by birth-death process methods. These values are in parentheses. In all cases, the 90% confidence intervals obtained surround the theoretical value.

Results of simulation for characteristics of the passage time P are in Tables 5.2 and 5.3. For N=2 jobs, with $\mu_0=1$, $\mu_1=0.5$ and $\psi=0.75$, these tables give an indication of the effect of different return states. The results are for two simulation runs, using the return states (1,1) and (2,1), respectively. Thus, for example, 2367 transitions in the Markov chain were required for 100 cycles of returns to the state (1,1) but only 1183 transitions for 100 cycles of returns to the state (2,1). Since returns to the state (2,1) occur approximately twice as frequently as to the state (1,1), we would expect that only half as many cycles (as for return state (1,1)) for comparable accuracy; this is borne out by the results in Tables 2 and 3. In all cases, the 90% confidence intervals obtained surround the theoretical value.

Results obtained for the response time R by simulation of the DL/I component model appear in Table 5.4. The initial state for the Markov chain $\underset{\sim}{X}$ is the state (1,N-2,0,0,0,0,0,6,N). Theoretical values for the α and β center utilizations along with the expected response time are obtainable from the analysis of the DL/I component model given by Lavenberg and Shedler (1976). Comparison of Table 5.4 with Table 5.1 reveals the effect on simulation running time of the considerable structural complexity of the DL/I component model; for the return states chosen, there are approximately the same number of transitions in the Markov chain for 250 cycles of the DL/I component model as for 400 cycles of the simple model. As before, the 90% confidence intervals for the expected response time surround the theoretical values.

TABLE 5.1

Simulation Results for Response Time R in Closed Network of Queues.
$N=2$, $\mu_0=1$, $\mu_1=0.5$, $p=0.75$. Return State is (1,1).

	No. of Cycles					
	100	200	400	800	1000	2000
Total simulated time	2606.44	5323.05	11647.92	23010.20	28541.88	57213.44
Fraction of time center 1 busy (0.8571)	0.8498	0.8483	0.8551	0.8498	0.8478	0.8544
Fraction of time center 2 busy (0.4286)	0.4237	0.4280	0.4171	0.4333	0.4384	0.4272
No. of transitions/cycle (M.C.)	22.56	22.81	25.11	24.59	24.29	24.59
No. of transitions/cycle (M.R.P.)	2.88	2.87	3.08	3.10	3.08	3.04
$E\{R\}$ (9.333)	9.050 ±0.6608	9.274 ±0.4496	9.447 ±0.3332	9.271 ±0.2250	9.282 ±0.1993	9.402 ±0.1476
$P\{R \le 2.333\}$	0.0799 ±0.0254	0.0714 ±0.0174	0.0714 ±0.0123	0.0725 ±0.0087	0.0719 ±0.0077	0.0721 ±0.0054
$P\{R \le 4.667\}$	0.2813 ±0.0436	0.2596 ±0.0296	0.2620 ±0.0213	0.2579 ±0.0149	0.2540 ±0.0131	0.2572 ±0.0090
$P\{R \le 9.333\}$	0.6076 ±0.0481	0.5993 ±0.0324	0.6067 ±0.0225	0.6201 ±0.0161	0.6195 ±0.0148	0.6174 ±0.0103
$P\{R \le 14.0\}$	0.8021 ±0.0407	0.8136 ±0.0276	0.8013 ±0.0143	0.8110 ±0.0131	0.8120 ±0.0117	0.8084 ±0.0083
$P\{R \le 18.667\}$	0.9271 ±0.0240	0.9129 ±0.0196	0.8970 ±0.0143	0.9081 ±0.0095	0.9073 ±0.0086	0.9012 ±0.0064

TABLE 5.2

Simulation Results for Passage Time P in Closed Network of Queues.
$N=2$, $\mu_0=1$, $\mu_1=0.5$, $p=0.75$. Return State is (1,1).

	No. of Cycles				
	100	200	400	800	1000
Total simulated time	2747.95	5463.94	11751.03	21413.70	27355.86
Fraction of time center 1 busy (0.8571)	0.8501	0.8465	0.8557	0.8507	0.8485
Fraction of time center 2 busy (0.4286)	0.4203	0.4305	0.4170	0.4312	0.4366
No. of transitions/cycle (M.C.)	23.67	23.31	25.33	22.97	23.27
No. of transitions/cycle (M.R.P.)	3.02	2.95	3.11	2.90	2.94
E{P} (6.667)	6.448 ±0.5668	6.594 ±0.3830	6.820 ±0.2895	6.570 ±0.2178	6.584 ±0.1950
P{P ≤ 1.667}	0.2119 ±0.0360	0.2105 ±0.0268	0.2068 ±0.0192	0.2180 ±0.0134	0.2122 ±0.0123
P{P ≤ 3.333}	0.4073 ±0.0410	0.3887 ±0.0306	0.3773 ±0.0212	0.3878 ±0.0161	0.3826 ±0.0142
P{P ≤ 6.667}	0.6457 ±0.0419	0.6180 ±0.0308	0.6138 ±0.0203	0.6290 ±0.0157	0.6317 ±0.0143
P{P ≤ 10.000}	0.8013 ±0.0346	0.7843 ±0.0244	0.7699 ±0.0179	0.7833 ±0.0137	0.7830 ±0.0125
P{P ≤ 13.333}	0.8940 ±0.0282	0.8879 ±0.0196	0.8656 ±0.0156	0.8733 ±0.0126	0.8736 ±0.0103

TABLE 5.3

Simulation Results for Passage Time P in Closed Network of Queues.
$N=2$, $\mu_0=1$, $\mu_1=0.5$, $p=0.75$. Return State is (2,1).

	No. of Cycles				
	100	200	400	800	1000
Total simulated time	1333.29	2727.80	5730.14	11074.78	13789.08
Fraction of time center 1 busy (0.8571)	0.8511	0.8537	0.8483	0.8563	0.8563
Fraction of time center 2 busy (0.4286)	0.4134	0.4161	0.4272	0.4163	0.4193
No. of transitions/cycle (M.C.)	11.83	11.75	12.21	11.94	11.87
No. of transitions/cycle (M.R.P.)	1.52	1.49	1.52	1.47	1.47
E{P} (6.667)	6.414 ±0.6865	6.656 ±0.6144	6.723 ±0.4474	6.754 ±0.3233	6.739 ±0.2852
P{P ≤ 1.667}	0.1974 ±0.0537	0.1919 ±0.0370	0.1987 ±0.0278	0.2044 ±0.0203	0.2029 ±0.0180
P{P ≤ 3.333}	0.3618 ±0.0615	0.3434 ±0.0447	0.3711 ±0.0331	0.3774 ±0.0239	0.3778 ±0.0210
P{P ≤ 6.667}	0.5855 ±0.0621	0.5993 ±0.0463	0.6125 ±0.0324	0.6234 ±0.0232	0.6290 ±0.0206
P{P ≤ 10.000}	0.7829 ±0.0544	0.7980 ±0.0411	0.7833 ±0.0288	0.7863 ±0.0205	0.7835 ±0.0182
P{P ≤ 13.337}	0.8816 ±0.0399	0.8822 ±0.0305	0.8686 ±0.0422	0.8634 ±0.0170	0.8625 ±0.0182

TABLE 5.4

Simulation Results for Response Time R in DL/I Component Model.
$N=2$, $E(\alpha_0)=6.7$, $E(\alpha_1)=33$, $E(\alpha_2)=1.5$, $E(\alpha_3)=1$, $E(\beta)=50$, $p_1=0.1$, $p_2=0.2$. Return State is (1,0,0,0,0,0,0,5,2).

	No. of Cycles					
	25	50	100	200	250	300
Total simulated time	3707.05	6061.72	14054.86	20771.41	35803.24	42669.02
Fraction of time center 1 busy (0.7498)	0.7485	0.8351	0.7640	0.7441	0.7404	0.7577
Fraction of time center 2 busy (0.5913)	0.6247	0.5435	0.6072	0.6147	0.6132	0.5966
No. of transitions/cycle (M.C.)	42.76	38.46	40.95	43.00	41.04	41.69
No. of transitions/cycle (M.R.P.)	1.44	1.38	1.47	1.67	1.63	1.70
E{R} (84.556)	102.974 ±24.638	87.851 ±16.533	95.611 ±19.237	89.136 ±11.837	87.969 ±10.125	83.501 ±8.657
P{R ≤ 75}	0.5555 ±0.1152	0.5942 ±0.0855	0.5918 ±0.0817	0.6198 ±0.0474	0.6241 ±0.0416	0.6399 ±0.0359
P{R ≤ 100}	0.5555 ±0.1152	0.6522 ±0.0871	0.6667 ±0.0790	0.6916 ±0.0487	0.6929 ±0.0420	0.7182 ±0.0367
P{R ≤ 125}	0.7222 ±0.1139	0.7681 ±0.0746	0.7483 ±0.0736	0.7485 ±0.0466	0.7543 ±0.0399	0.7730 ±0.0343
P{R ≤ 150}	0.7778 ±0.1094	0.8261 ±0.0606	0.8163 ±0.0623	0.8024 ±0.0440	0.8034 ±0.0382	0.8024 ±0.0391
P{R ≤ 175}	0.8333 ±0.0859	0.8551 ±0.0557	0.8435 ±0.0562	0.8413 ±0.0400	0.8403 ±0.0355	0.8413 ±0.0309

6.0. FINITE CAPACITY OPEN NETWORKS OF QUEUES

We now consider open networks of service centers with jobs arriving at the network, traversing the network and receiving service along the way, and finally departing from the network. The network structure we permit is essentially the same as that described in Section 3.0, except that here the networks are open; thus arrivals from an external source and departures to an external sink occur. Only a finite number of jobs, N, may reside in the network at a given time. We consider two formulations of finite capacity open networks. In "arrival process shutdown" models, a job arriving when the network already contains N-1 customers causes the arrival process to shut down; the arrival process remains shut down until the first subsequent departure from the network. In "jobs turned away" models, the arrival process never shuts down, but jobs arriving when the network already contains N jobs turn away.

6.1. Markov Arrival Processes

The arrival processes we allow are particular stochastic point processes (series of events) associated with time-homogeneous Markov renewal processes. Let J be a finite or countable set, $\underset{\sim}{W}=\{W_n:n\geq 0\}$ random variables assuming values in J, and $\underset{\sim}{U}=\{U_n:n\geq 0\}$ nonnegative random variables such that $0=U_0\leq U_1\leq U_2\leq\ldots$. Recall that a stochastic process $(\underset{\sim}{W},\underset{\sim}{U})=\{(W_n,U_n):n\geq 0\}$ is a Markov renewal process provided that

$$P\{W_{n+1} = j,\ U_{n+1}-U_n\leq t\,|\,W_0,\ldots,W_n;\ U_0,\ldots,U_n\}$$

$$= P\{W_{n+1}=j,\ U_{n+1}-U_n\leq t\,|\,W_n\}\ ,$$

with probability one for all $n\geq 0$, $j\epsilon J$, and $t\geq 0$, and is time-homogeneous provided that $P\{W_{n+1}=j, U_{n+1}-U_n\leq t|W_n\}$ is independent of n. We denote by $\underset{\sim}{M}=\{M(t):t\geq 0\}$ the semi-Markov process generated by $(\underset{\sim}{W},\underset{\sim}{U})$, i.e.,

$$M(t) = W_n, \text{ if } U_n \leq t < U_{n+1} . \tag{6.1.1}$$

By definition, a <u>Markov arrival process</u> is a stochastic point process $\underset{\sim}{U}$ that satisfies the condition

$$P\{U_{n+1}-U_n\leq t|W_n=i\} = 1 - \exp(-\lambda_i t) \tag{6.1.2}$$

for all $i\epsilon J$ and $t\geq 0$, where $\lambda_i>0$. Under this condition the semi-Markov process $\underset{\sim}{M}$ is equivalent to a continuous time Markov chain. Throughout, we assume that $\underset{\sim}{M}$ is irreducible and positive recurrent, and that the Markov renewal process $(\underset{\sim}{W},\underset{\sim}{U})$ is independent of the service times and Markov routing of jobs in the network. Note that in "arrival process shutdown" models we can think of the arrival process as operating in virtual time, i.e., the nth customer arrives at virtual time U_n. The actual time of the nth arrival, however, may be somewhat later due to the finite capacity constraint. In "jobs turned away" models, the nth job arrives at time U_n, but may be turned away.

(6.1.3) EXAMPLE. <u>Poisson Process</u>. Let $J=\{1\}$ so that $W_n=1$ with probability one for all $n\geq 0$, and for $t\geq 0$ let

$$P\{U_{n+1}-U_n\leq t\} = 1 - \exp(-\lambda t) ,$$

where $\lambda>0$. Clearly, the Markov arrival process $\underset{\sim}{U}=\{U_n:n\geq 0\}$ is a Poisson process.

(6.1.4) EXAMPLE. Switching Poisson Process. Let $J=\{1,2,...,k\}$, $q_{ij}=P\{W_{n+1}=j|W_n=i\}$ for $i,j\epsilon J$, and for $t\geq 0$ let

$$P\{U_{n+1}-U_n\leq t|W_n=i\} = 1 - \exp(-\lambda_i t) \quad ,$$

where $\lambda_i>0$ for $i=1,2,...,k$. Here the successive times-between-events are exponentially distributed with parameters which are governed by the transition matrix $\{q_{ij}:i,j\epsilon J\}$. This Markov arrival process is a semi-Markov generated point process with exponential holding times.

(6.1.5) EXAMPLE. Branching Poisson Process. The branching Poisson process is a model for clustered arrivals. Consider a particular branching Poisson process constructed as follows. A Poisson process with parameter λ_1 generates a series of primary events, and with independent probability $r\epsilon(0,1]$, a primary event initiates a subsidiary series of events. Each subsidiary process consists of a geometrically distributed number of subsidiary events, and the times between these subsidiary events are independent and exponentially distributed with parameter λ_2. Finally, the branching Poisson process is the superposition of all primary and subsidiary events. To represent this process as a Markov arrival process, we set $J=\{0,1,2,...\}$ and identify W_n with the number of subsidiary processes active at the time of the nth event of the branching Poisson process. Let $p\epsilon(0,1)$ be the parameter associated with the geometric distribution (mean $1/(1-p)$) governing the number of subsidiary events in each subsidiary process; that is, the probability of k subsidiary events in a given subsidiary process is $p^{k-1}q$ $(k\geq 1)$, where $q=1-p$. Then for $i\epsilon J$,

$$P\{W_{n+1}=j \mid W_n=i\} = \begin{cases} i\lambda_2 q/(\lambda_1+i\lambda_2) & , \quad j=i-1 \\ [i\lambda_2 p+(1-r)\lambda_1]/(\lambda_1+i\lambda_2), & \quad j=i \\ r\lambda_1/(\lambda_1+i\lambda_2) & , \quad j=i+1 , \end{cases}$$

and for $t \geq 0$,

$$P\{U_{n+1}-U_n \leq t \mid W_n=i\} = 1 - \exp[-(\lambda_1+i\lambda_2)t] .$$

6.2. Networks of Queues and Associated Stochastic Processes

As before, we permit a finite number of single-server service centers, s, and a finite number of (mutually exclusive) job classes, c. At every time epoch each job is in exactly one job class, but jobs may change class as they traverse the network. Upon completion of service at center i, a job of class j goes to center k and changes to class ℓ with probability $p_{ij,k\ell}$, (where $1 \leq i,k \leq s$ and $1 \leq j,\ell \leq c$) and $\underline{P}=\{p_{ij,k\ell}\}$. A job from the external source arrives at center k as a job of class ℓ with probability $p_{k\ell}$, and a job of class ℓ completing service at center k departs to the external sink with probability $q_{k\ell}$.

At each service center jobs queue and receive service according to a fixed priority scheme among classes; the priority scheme may differ from center to center. Within a class at a center, jobs receive service according to a fixed discipline, and, some centers may never see jobs of certain classes as determined by the routing matrix $\underline{P}$. A job in service at a center may or may not be preempted if another job of higher priority joins the queue at the center. All service times in the network are mutually independent, and at a particular center have a distribution consisting of exponential stages with parameters which may depend on the service center, class of job being served, and "state" of the entire network.

As in Section 3.2, we let $S_i(t)$, $i=1,2,\ldots,s$, denote the class of job receiving service at center i at time t, with $S_i(t)=0$ if at time t there are no jobs at center i. The classes of jobs serviced at center i ordered by decreasing priority are $j_1(i),j_2(i),\ldots,j_{k(i)}(i)$, all elements of the set $\{1,2,\ldots,c\}$. Denote by $C_{j_1}^{(i)}(t),C_{j_2}^{(i)}(t),\ldots,C_{j_{k(i)}}^{(i)}(t)$ the number of jobs in queue at time t of the various classes of jobs served at center i.

To deal with passage times, we again use the idea of the position of a distinguished "marked" job. We continue to think of the (at most N) jobs in the network ordered in a linear stack according to the following scheme, and define the vector Z(t) by

$$Z(t) = (C_{j_{k(1)}}^{(1)}(t),C_{j_{k(1)-1}}^{(1)}(t),\ldots,C_{j_1}^{(1)}(t),S_1(t);\ldots;$$

$$C_{j_{k(s)}}^{(s)}(t),C_{j_{k(s)-1}}^{(s)}(t),\ldots,C_{j_1}^{(s)}(t),S_s(t)) .$$

The linear job stack corresponds to the order of components in the vector Z(t) after ignoring any zero components. Within a class at a particular service center, jobs waiting appear in the job stack in FCFS order, the latest to arrive being closest to the top of the stack. We again denote by N(t) the position (from the top) of the marked job in this job stack at time t.

Recalling that $\underset{\sim}{M}$ is the semi-Markov process of Equation (6.1.1) associated with the Markov arrival process, for $t \geq 0$ we specify the state of the network by

$$X(t) = (M(t), Z(t), N(t), U(t)) \ ,$$

Here $U(t)=(U_1(t),\ldots,U_N(t))$ and $U_j(t)$ $(1 \leq j \leq N)$ is the stage of service associated with the jth job in the job stack; $U_j(t)=0$ if there are less than j jobs in the system at time t. By virtue of the service time distributional assumptions, the Markovian routing structure, and the definition of a Markov arrival process, $\underset{\sim}{X}=\{X(t):t \geq 0\}$ is a continuous time Markov chain with a (possibly countable) state space E.

6.3. Job Marking

The principal concern here remains the estimation of general characteristics of passage times, the times required for a job to traverse a specified portion of the open network. In a finite capacity open network, a passage time is termed a "response time" if it is the total time a job is in the network. To estimate passage times, we track an appropriate sequence of typical jobs, based on the idea of a marked job, and measure the passage times for a sequence of marked jobs; these are to be typical jobs in the sense that the sequence of passage times for the marked jobs should converge in distribution to the same random variables as do the passage times for all jobs. It is necessary to take some care to ensure that this is the case.

Our *job marking scheme* is as follows. Let i $(0 \le i < N)$ be the number of jobs left behind by the marked job at the epoch at which it departs from the network (i.e., goes to the external sink). Then for "arrival process shutdown" models, the $(N-i)$th subsequent arrival is the next marked job. For "jobs turned away" models, the $(N+1-i)$th subsequent arrival is the next marked job.

Note that this marking scheme ensures that there is at most one marked job in the network at a time, and thus there is no need for further augmentation of the numbers-in-queue, stages-of-service state vector for the measurement of passage times. (If there is no marked job in the network at time t, we define the position N(t) in the linear job stack to be zero.) Note also that more than one passage time can start (and terminate) for a particular marked job before it departs from the network.

To see that the sequence of passage times for jobs marked by this scheme has the desired property, consider "arrival process shutdown" models. We introduce a so-called "phantom server" which generates the times $\underset{\sim}{U}=\{U_n : n \ge 0\}$ according to the Markov renewal process $(\underset{\sim}{W},\underset{\sim}{U})$ of the Markov arrival process. Assuming that i $(0 \le i < N)$ is the number of jobs left behind by the marked job at the instant it departs from the network, we route the marked job to the queue at the phantom server where $N-i-1$ jobs are already residing. Upon completion of service to a job by the phantom server, the job returns to the network in the same manner as arriving jobs, i.e., with probability $p_{k\ell}$ the job goes to center k and becomes class ℓ. This method generates arrivals to the network in exactly same way the original arrival process does with the finite capacity constraint, namely, arrivals occur at the times U_n provided the number in the network is less than N. Figure 6.1 illustrates the general flow of jobs. It is now clear that in effect we have a closed network of queues in which the marked job never leaves the network. Furthermore, the stochastic structure of the original problem remains. The chief advantage of this device is that we can use the simulation method developed in Section 3 for closed networks. Note that there is no need for a "$\rho<1$" condition to guarantee stability of the open network since it is effectively a closed network with a finite number of jobs.

For "jobs turned away" models, we introduce a phantom server in a similar way; see Figure 6.2. A job completing service at the phantom server with j-1 other jobs at the server returns to the network with probability 1-p(j), where p(j)=1 if j=1 and 0 otherwise. We are in effect considering a closed network with N+1 jobs. It is straightforward to check that the simulation method of Section 4.1 extends to closed networks in which, as here, routing probabilities may depend on the number of jobs in the service center.

Having reduced the problem to estimation in a closed network, it is necessary to modify the Markov chain $\underset{\sim}{X}$. We view the phantom server as service center s+1, serving only one class of jobs. Let $Q_{s+1}(t)$ denote the total number of jobs at center s+1 at time t. Then we augment the vector Z(t) with the component $Q_{s+1}(t)$, and define X(t) as before but with this augmented Z(t). We also modify the Markov routing matrix $\underline{P}=\{p_{ij,k\ell}\}$

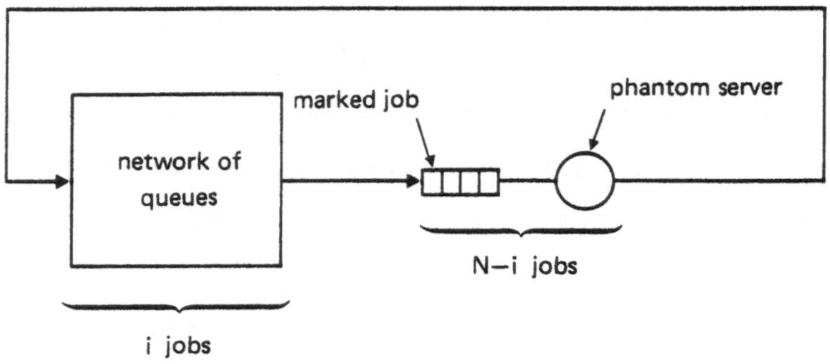

Figure 6.1. Flow of jobs with phantom server added. "Arrival process shutdown" model

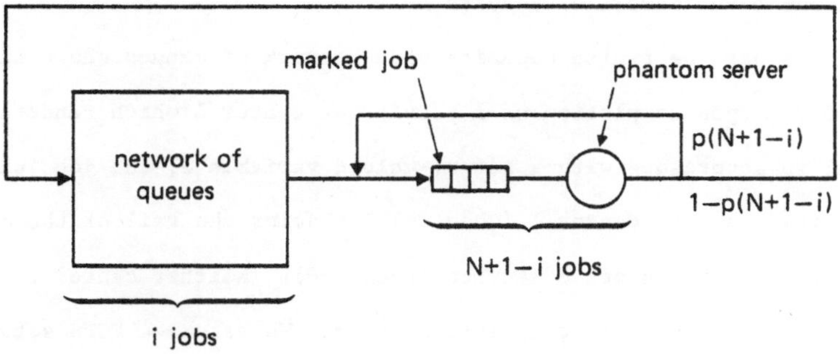

Figure 6.2. Flow of jobs with phantom server added. "Jobs turned away" model

to describe the closed network, and assume that the resulting matrix is irreducible. The assumptions made on the network imply that the Markov chain $\underset{\sim}{X}$ is irreducible and positive recurrent, and it is possible to proceed with the estimation of passage times as before.

6.4. Example and Numerical Results

We consider the finite capacity open network of queues shown in Figure 6.3. Upon completion of a service at center 1 which renders α service, in accordance with a binary-valued variable ψ, the job joins the tail of the queue in center 1 (when $\psi=1$) or joins the tail of the queue in center 2 which renders β service (when $\psi=0$). Neither center 1 nor center 2 service is subject to interruption. We assume a FCFS service discipline in each queue. The limiting response time of interest (denoted by R) is the time measured from arrival of a job at the center 1 queue until departure of the job. Also of possible interest in this model is the limiting passage time (denoted by P) defined as the time measured from arrival of a job into the center 1 queue until the job enters the center 2 queue.

For each of the arrival processes described below, we consider simulation of this network of queues under the following probabilistic assumptions:

(i) successive α service times form a sequence of i.i.d. random variables, exponentially distributed with rate parameter μ_0;

(ii) successive β service times form a sequence of i.i.d. random variables exponentially distributed with rate parameter μ_1;

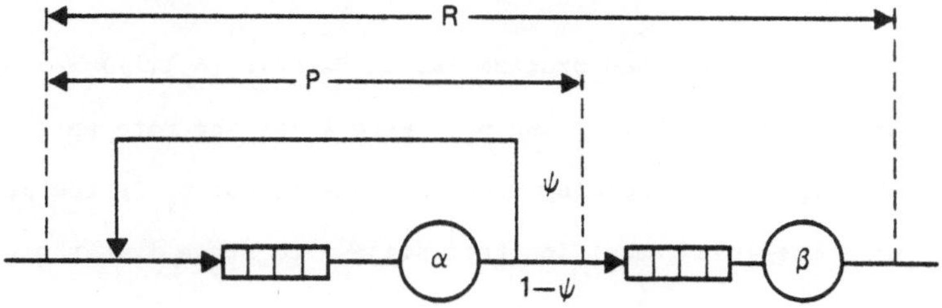

(i) Service center 1 renders α service

(ii) Service center 2 renders β service

(iii) α and β services are not interruptable

(iv) Routing determined by binary valued random variable ψ

Figure 6.3. Finite capacity open network of queues

(iii) ψ is a Bernoulli random variable and values of ψ at successive α service completions form a sequence of i.i.d. random variables;

(iv) the sequences in (i), (ii) and (iii) are mutually independent.

For the simulation, the generator of arrivals is either

(i) a Poisson process (as in Example (6.1.3)) having rate parameter λ;

or

(ii) a switching Poisson process (as in Example (6.1.4)) having parameters λ_1, λ_2, p_1 and p_2. Here λ_i is the rate parameter of the exponential holding time in state i, and p_i is the probability of a one-step transition from state i to state i in the embedded Markov chain.

In this network, there are two classes of jobs, class 1 jobs at center 1 and class 2 jobs at center 2. Since each center sees only one class of job, we can define

$$Z(t) = (Q_1(t), Q_2(t)) ,$$

where

$Q_1(t)$ = number of jobs waiting or in service at center 1 at time t ,

and

$Q_2(t)$ = number of jobs waiting or in service at center 2 at time t .

As above, let N(t) denote the position of the marked job in the job stack corresponding to the order of the nonzero components of Z(t), and let M(t) denote the semi-Markov process associated with the Markov arrival process. Then, since the service times are exponentially distributed, the process $\underset{\sim}{X}=\{X(t):t\geq 0\}$, where

$$X(t) = (M(t),Q_1(t),Q_2(t),N(t)) ,$$

is the underlying continuous time Markov chain.

For this model the state space E of $\underset{\sim}{X}$ is

$$E = \{(m,i,j,k):0\leq i,j,k\leq N;\ i+j\leq N;\ m\in J\} ,$$

where N is the maximum number of jobs in the network.

The Markov chain governing change of job class can be taken to have state space {1,2} and transition matrix

$$\underline{P} = \begin{bmatrix} p & 1-p \\ 0 & 0 \end{bmatrix},$$

where we denote by p the probability that the branching variable ψ takes the value 1. The probabilities $p_{k\ell}$ governing routing of jobs from the external source are $p_{11}=1$, $p_{12}=0$, $p_{21}=0$, $p_{22}=0$, and the probabilities $q_{k\ell}$ of departure to the external sink are $q_{11}=0$, $q_{22}=0$, $q_{21}=0$, $q_{22}=1$. The 3×3 matrix governing change of job classes for the associated closed network is

$$\begin{bmatrix} p & 1-p & 0 \\ 0 & 0 & 1 \\ 1 & 0 & 0 \end{bmatrix}.$$

The sets A_1 and A_2 defining the start of response times for the marked job are

$$A_1 = \{(m,i,j,k)\in E: k=N\}$$

and

$$A_2 = \{(m,i,j,1)\in E: i>0\} .$$

The sets B_1 and B_2 defining the termination of response times for the marked job are

$$B_1 = \{(m,i,j,k)\in E: k=i+j,\ j>0\}$$

and

$$B_2 = \{(m,i,j,k)\in E: k=i+j+1\} .$$

Tables 6.1-6.3 contain results obtained for the response time R by simulation of the network of Figure 6.3 for N=4 jobs with $\mu_0=1$, $\mu_1=0.5$ and p=0.75. The initial state for the Markov chain $\underset{\sim}{X}$ (and return state identifying cycles) is the state (1,4,0,1). Simulation results for arrivals generated by a Poisson process of rate $\lambda=0.4$ appear in Table 6.1. For Poisson arrivals the two formulations of finite capacity networks are equivalent. Theoretical values for the long-run expected fractions of time that centers 1 and 2 are busy (utilizations), and the expected response time, obtainable either by birth-death or semi-Markov process methods, appear in parentheses. Note that with the exception of the 1000 cycle run, the 90% confidence intervals obtained surround the theoretical value. The percentile points are, respectively, 0.25, 0.5, 1, 1.5, and 2 times the theoretical mean response time.

Table 6.2 gives results of simulation of the "arrival process shutdown" model for arrivals generated by a two-state switching Poisson process with $\lambda_1=2$, $\lambda_2=0.667$, $p_1=0.978$ and $p_2=0.865$. With these parameter values, events in the stationary switching Poisson process occur at rate 0.4, and the stationary times-between-events have coefficient of variation equal to 2 and serial correlation coefficient of lag 1 (cf. Cox and Lewis (1966), p. 196) equal to 0.375. In all cases, the 90% confidence intervals for $E\{R\}$ surround the theoretical values. Corresponding results for the "jobs turned away" model are in Table 6.3. Again, the 90% confidence intervals for $E\{R\}$ surround the theoretical value.

An overall observation from Tables 6.1-6.3 is that the lengths of confidence intervals obtained (for expected response time as well as percentiles of response time) from the three simulations are roughly comparable.

TABLE 6.1

Simulation Results for Response Time R in Finite Capacity Open Network of Queues. Arrivals Generated by Poisson Process with $\lambda=0.4$, $N=4$, $\mu_0=1$, $\mu_1=0.5$, $p=0.75$.

	No. of Cycles					
	100	200	400	800	1000	2000
Total simulated time	6841.47	13203.06	26584.25	55647.25	69936.01	138199.66
Fraction of time center 1 busy (0.8810)	0.8709	0.8853	0.8794	0.8809	0.8818	0.8812
Fraction of time center 2 busy (0.4405)	0.4374	0.4356	0.4416	0.4419	0.4419	0.4423
No. of transitions/cycle (M.C.)	71.45	70.58	70.24	73.77	74.18	73.55
No. of transitions/cycle (M.R.P.)	3.66	3.57	3.84	3.78	3.80	3.78
$E\{R\}$ (13.687)	14.438 ±0.897	14.289 ±0.667	13.895 ±0.452	13.978 ±0.310	14.007 ±0.285	13.818 ±0.202
$P\{R \leq 3.422\}$	0.0929 ±0.0265	0.0842 ±0.0172	0.0824 ±0.0118	0.0871 ±0.0082	0.0875 ±0.0073	0.0854 ±0.0052
$P\{R \leq 6.844\}$	0.2623 ±0.0338	0.2833 ±0.0251	0.2841 ±0.0173	0.2916 ±0.0124	0.2964 ±0.0111	0.2939 ±0.0083
$P\{R \leq 13.687\}$	0.6202 ±0.0377	0.6171 ±0.0281	0.6239 ±0.0206	0.6124 ±0.0139	0.6150 ±0.0124	0.6182 ±0.0089
$P\{R \leq 20.531\}$	0.7650 ±0.0333	0.7770 ±0.0244	0.7893 ±0.0174	0.7941 ±0.0114	0.7947 ±0.0103	0.7996 ±0.0073
$P\{R \leq 27.375\}$	0.8552 ±0.0304	0.8710 ±0.0211	0.8868 ±0.0133	0.8908 ±0.0088	0.8875 ±0.0082	0.8928 ±0.0058

TABLE 6.2

Simulation Results for Response Time R in Finite Capacity Open Network of Queues. Arrivals Generated by Two-state Switching Poisson Process with λ_1=2, λ_2=0.0667, p_1=0.978, p_2=0.865, N=4, μ_0=1, μ_1=0.5, p=0.75. "Arrival Process Shutdown" Model.

	No. of Cycles					
	100	200	400	800	1000	2000
Total simulated time	6025.55	12182.34	26376.79	47896.37	59014.54	115393.31
Fraction of time center 1 busy (0.7129)	0.6786	0.6718	0.6518	0.6910	0.6988	0.7009
Fraction of time center 2 busy (0.3565)	0.3311	0.3281	0.3251	0.3502	0.3528	0.3519
No. of transitions/cycle (M.C.)	50.61	51.07	53.44	51.85	51.89	51.08
No. of transitions/cycle (M.R.P.)	2.52	2.49	2.61	2.57	2.56	2.51
E{R} (14.897)	14.935 ±1.120	15.120 ±0.944	14.971 ±0.707	14.843 ±0.460	14.943 ±0.408	14.983 ±0.285
P{R ≤ 3.724}	0.0833 ±0.0305	0.0743 ±0.0190	0.0931 ±0.0167	0.0850 ±0.0118	0.0863 ±0.0103	0.0848 ±0.0069
P{R ≤ 7.448}	0.2897 ±0.0489	0.2952 ±0.0344	0.3042 ±0.0269	0.2904 ±0.0190	0.2861 ±0.0164	0.2831 ±0.0135
P{R ≤ 14.897}	0.6032 ±0.0488	0.6185 ±0.0391	0.6142 ±0.0293	0.6221 ±0.0195	0.6155 ±0.0172	0.6205 ±0.0117
P{R ≤ 22.345}	0.7897 ±0.0389	0.7871 ±0.0314	0.7927 ±0.0224	0.8009 ±0.0152	0.7970 ±0.0135	0.7985 ±0.0095
P{R ≤ 29.793}	0.8810 ±0.0318	0.8775 ±0.0246	0.8820 ±0.0164	0.8951 ±0.0111	0.8934 ±0.0099	0.8915 ±0.0084

TABLE 6.3

Simulation Results for Response Time R in Finite Capacity Open Network of Queues. Arrivals Generated by Two-state Switching Poisson Process with $\lambda_1=2$, $\lambda_2=0.667$, $p_1=0.978$, $p_2=0.865$, $N=4$, $\mu_0=1$, $\mu_1=0.5$, $p=0.75$. "Jobs Turned Away" Model.

	No. of Cycles				
	100	200	400	800	1000
Total simulated time	22753.44	46393.70	78702.29	162477.07	194018.01
Fraction of time center 1 busy (0.4289)	0.3941	0.4110	0.4297	0.4331	0.4360
Fraction of time center 2 busy (0.2144)	0.1993	0.2090	0.2161	0.2178	0.2185
No. of transitions/cycle (M.C.)	93.27	98.37	87.51	90.71	87.37
No. of transitions/cycle (M.R.P.)	4.50	4.73	4.21	4.38	4.23
E{R} (11.928)	11.753 ±0.666	11.949 ±0.475	12.029 ±0.381	12.002 ±0.279	11.997 ±0.257
P{R ≤ 2.982}	0.1333 ±0.0288	0.1078 ±0.0189	0.0955 ±0.0134	0.1068 ±0.0089	0.1097 ±0.0082
P{R ≤ 5.964}	0.3111 ±0.0366	0.2875 ±0.0254	0.2849 ±0.0186	0.3088 ±0.0136	0.3114 ±0.0122
P{R ≤ 11.928}	0.6333 ±0.0393	0.6332 ±0.0261	0.6249 ±0.0195	0.6258 ±0.0139	0.6294 ±0.0127
P{R ≤ 17.892}	0.7933 ±0.0305	0.8034 ±0.0216	0.8006 ±0.0170	0.7949 ±0.0114	0.7971 ±0.0103
P{R ≤ 23.856}	0.8760 ±0.0218	0.8953 ±0.0148	0.8979 ±0.0121	0.8900 ±0.0086	0.8898 ±0.0078

7.0. MARKED JOB SIMULATION VIA HITTING TIMES

In this section we develop a new stochastic setting for the marked job method based on "hitting times" of an underlying continuous time Markov chain to fixed sets of states. This formulation is the basis for the subsequent discussion. We first make some remarks about the results obtained in the previous sections.

The earlier discussion of the marked job method applies to networks of queues having single server service centers. The generalization to networks having multiple server service centers, however, is straightforward. To handle these networks, it is sufficient to incorporate into the job stack (and associated state vector) information which gives the class of job being serviced by each of the servers at a multiple server service center.

A further generalization to networks with stochastically nonidentical jobs is also possible. The marked job method of Section 4 applies equally well to networks in which jobs are of a finite number of types. In such networks, jobs of each type have their own routing structures and service requirements, and in the case of finite capacity open networks, independent arrival processes. We return to this topic in Section 10.

Although the finite capacity constraint on the open queueing networks considered in Section 6 appears natural in many modelling contexts, it would be of interest to extend the marked job method to open networks of infinite capacity. The main barrier to doing so is the definition of a sequence of marked jobs whose nonoverlapping measured passage times converge in distribution to the same random variable as does the entire sequence of passage times (enumerated, say, in order of start times). This remains an open problem.

In the remainder of this section, we develop a stochastic setting for passage time simulation using a marked job which is simpler than that of Section 4.2; the definition of passage time, however, is somewhat more restrictive. Here, the starts of passage times for the marked job are the successive entrances (hitting times) of a particular continuous time Markov chain to a fixed set of states, and the terminations of such passage times are the successive entrances to another fixed set of states. This formulation provides the basis for the decomposition method for passage time simulation in Section 8, the analysis of statistical efficiency in Section 9, and the discussion in Section 10 of methods for networks with multiple job types.

7.1 Preliminaries

As in Section 3, we consider closed networks of queues with a finite number of jobs, N. In each network there are a finite number of service centers, s, and a finite number of job classes, c. At each epoch of time each job is in exactly one job class, but jobs may change class as they traverse the network. Upon completion of service at center i, a job of class j goes to center k and changes to class ℓ with probability $p_{ij,k\ell}$. We assume that $\underline{P}=\{p_{ij,k\ell}:1\le i,k\le s,\ 1\le j,\ell\le c\}$ is a given irreducible Markov matrix.

At each service center jobs queue and receive service according to a fixed priority scheme among classes, which scheme can vary from center to center. Each center operates as a single server, processing jobs of a fixed class according to a fixed service discipline. All service times in the network are mutually independent, and at each center have a distribution with a Cox-phase representation with parameters which may depend on the service center, class of job being served, and the "state" of the entire system. (Recall that we exclude zero service times occurring with positive probability.) A job in service may or may not be preempted (according to a fixed procedure for each center) if another job of higher priority joins the queue at the center. We restrict the present discussion to networks in which all service times are exponentially distributed, and deal with distributions having an Cox-phase representation in the usual way by the method of stages.

We review the notation of Section 3.2 used to characterize the state of the network at time t. For $i=1,2,\ldots,s$, $S_i(t)$ denotes the class of the job receiving service at center i at time t; $S_i(t)=0$ if there are no jobs at center i at time t. We denote by $j_1(i),\ldots,j_{k(i)}(i)$ the job classes served at center i, ordered by decreasing priority, and $C_{j_1}^{(i)}(t),\ldots,C_{j_{k(i)}}^{(i)}(t)$ denote the number of jobs in queue at time t of the various classes served at center i.

As in Section 3.2, we view the N jobs as being completely ordered in a linear job stack, and define the vector Z(t) according to

$$Z(t) = (C^{(1)}_{j_{k(1)}}(t),\ldots,C^{(1)}_{j_1}(t),S_1(t);\ldots;$$
$$C^{(s)}_{j_{k(s)}}(t),\ldots,C^{(s)}_{j_1}(t),S_s(t)) \; . \tag{7.1.1}$$

The job stack corresponds to the order of components in the vector Z(t) after ignoring any zero components. Within a class at a center, jobs waiting appear in the job stack in the order of their arrival at the center, the latest to arrive being closest to the top of the stack.

Letting N(t) denote the position from the top of the marked job in this job stack at time t, for $t\geq 0$ the state vector of the network is

$$X(t) = (Z(t),N(t)) \; . \tag{7.1.2}$$

As before, we specify the passage time for the marked job by four subsets (A_1, A_2, B_1, and B_2) of E, the state space of the process $\underset{\sim}{X}=\{X(t):t\geq 0\}$. The sets A_1 and A_2 [respectively B_1 and B_2] determine when to start [respectively stop] the clock measuring a particular passage time for the marked job. Denoting the jump times of $\underset{\sim}{X}$ by $\{\tau_n:n\geq 0\}$, for $k,n\geq 1$ we require that the sets A_1, A_2, B_1 and B_2 satisfy:

if $X(\tau_{n-1})\in A_1$, $X(\tau_n)\in A_2$, $X(\tau_{n-1+k})\in A_1$ and $X(\tau_{n+k})\in A_2$, then $X(\tau_{n-1+m})\in B_1$ and $X(\tau_{n+m})\in B_2$ for some $0<m\leq k$;

and

if $X(\tau_{n-1})\in B_1$, $X(\tau_n)\in B_2$, $X(\tau_{n-1+k})\in B_1$ and $X(\tau_{n+k})\in B_2$, then $X(\tau_{n-1+m})\in A_1$ and $X(\tau_{n+m})\in A_2$ for some $0\leq m<k$.

Recall that these conditions ensure that the start and termination times for the specified passage time strictly alternate. Also in terms of these jump times, we define two sequences of random times: $\{S_j:j\geq 0\}$ and $\{T_j:j\geq 1\}$. The start [respectively termination] time of the jth passage time for the marked job is denoted by S_{j-1} [respectively T_j]. Assuming that a passage time for the marked job begins at $t=0$, we have

$$S_0 = 0$$

$$S_j = \inf\{\tau_n \geq T_j : X(\tau_n)\epsilon A_2,\ X(\tau_{n-1})\epsilon A_1\},\ j\geq 1$$

and

$$T_j = \inf\{\tau_n > S_{j-1} : X(\tau_n)\epsilon B_2,\ X(\tau_{n-1})\epsilon B_1\},\ j\geq 1\ . \qquad (7.1.3)$$

The jth passage time for the marked job is $P_j=T_j-S_{j-1}$, $j\geq 1$. Note that these definitions are more restrictive than those of Equation (4.2.2) in that the value of k in the earlier equation is set equal to n-1.

7.2. Simulation for Passage Times

We let X_n denote the state of the continuous time Markov chain $\underset{\sim}{X}$ when the (n+1)st passage time of the marked job begins: $X_n=X(S_n)$, $n\geq 0$. The process $\{X_n:n\geq 0\}$ is a discrete time Markov chain (with state space A_2) which we assume to be irreducible and aperiodic. By Theorem (4.2.5), the process $\{(X_n,P_{n+1}):n\geq 0\}$ is a regenerative process in discrete time, and the regenerative property guarantees that as $n\to\infty$,

$$(X_n,P_{n+1}) \Rightarrow (X,P)\ .$$

The random variable P is the limiting passage time for the marked job. The argument in Appendix 1 shows that the sequence of passage times for any other job also converges in distribution to the same random variable P. Moreover, the sequence of passage times (irrespective of job identity) enumerated according to start times (or termination times) also converges to P. The goal of the simulation is the estimation of characteristics of the limiting passage time P. Let f be a real-valued measurable function with domain R_+, and define the quantity r(f) according to

$$r(f) = E\{f(P)\} . \tag{7.2.1}$$

We now depart from the development of Section 4.2. Let L(t) denote the last state visited by the Markov chain $\underset{\sim}{X}$ before jumping to X(t), and define the stochastic process $\underset{\sim}{V}=\{V(t):t\geq 0\}$ by

$$V(t) = (L(t),X(t)) . \tag{7.2.2}$$

In this development of the marked job method, the process $\underset{\sim}{V}$ is the fundamental stochastic process of the passage time simulation.

The process $\underset{\sim}{V}$ has a state space F consisting of all pairs of states (i,j), $i,j\epsilon E$, for which a transition in $\underset{\sim}{X}$ from state i to state j can occur with positive probability. Since $\underset{\sim}{X}$ is an irreducible, positive recurrent Markov chain, so is $\underset{\sim}{V}$. The entrance times of $\underset{\sim}{V}$ to a state $(i,j)\epsilon F$ correspond to the times of transition in $\underset{\sim}{X}$ from state i to state j. We define subsets S and T of F according to

$$S = \{(k,m)\epsilon F:k\epsilon A_1,\ m\epsilon A_2\}$$

and

$$T = \{(k,m)\epsilon F:k\epsilon B_1,\ m\epsilon B_2\} , \tag{7.2.3}$$

and observe that the entrances of $\underset{\sim}{V}$ to S [respectively T] correspond to the starts [respectively terminations] of passage times for the marked job. In the case of response times, S=T.

The next step is to select a fixed element of S, which for convenience we designate state 0, and assume that V(0)=0. We let $\{V_n:n\geq 0\}$ denote the embedded jump chain associated with the continuous time process $\underset{\sim}{V}$. The random times $\{\gamma_n:n\geq 1\}$ denote the lengths of the successive 0-cycles (successive returns to the fixed state 0) for $\{V_n:n\geq 0\}$, and we define $\delta_0=0$ and $\delta_m=\gamma_1+\ldots+\gamma_m$, $m\geq 1$. Then the number of passage times for the marked job in the first 0-cycle of $\underset{\sim}{V}$ is

$$M_1 = \sum_{j=0}^{\delta_1-1} 1_{\{V_j\in S\}} .$$

(Recall that for a set A, $1_A(x)=1$ if $x\in A$ and 0 if $x\notin A$. Here we suppress the argument ω.) The sum of the values of the function f for the passage times for the marked job in this cycle is simply

$$Y_1(f) = \sum_{j=1}^{M_1} f(P_j) .$$

We denote the analogous quantities in the kth 0-cycle by M_k and $Y_k(f)$. Since $\underset{\sim}{V}$ is an irreducible, positive recurrent Markov chain, it is a regenerative process, and the pairs of random variables $\{(Y_k(f),M_k):k\geq 1\}$ are independent and identically distributed. Then, provided that $P\{P\in D(f)\}=0$, where D(f) is the set of discontinuities of the function f in the definition of r(f), and $E\{|f(P)|\}<\infty$, it follows that

$$r(f) = E\{f(P)\} = E\{Y_1(f)\}/E\{M_1\} .$$

Therefore, the regenerative method applies and (from a fixed number n of 0-cycles) provides the strongly consistent point estimate $\hat{r}_n(f)=\overline{Y}_n(f)/\overline{M}_n$ for r(f). Here $\overline{Y}_n(f)=(Y_1(f)+...+Y_n(f))/n$ and $\overline{M}_n=(M_1+...+M_n)/n$. The associated confidence interval is based on the central limit theorem

$$\frac{n^{1/2}\{\hat{r}_n(f)-r(f)\}}{\sigma/E\{M_1\}} \Rightarrow N(0,1) ,$$

where σ^2 is the variance of $Y_1(f)-r(f)M_1$. It is easy to check that these point and interval estimates for r(f) (obtained in the setting of the process $\underset{\sim}{V}$) are the same as those obtained from Algorithm (4.1.1).

(7.2.4) EXAMPLE. Recall the model of Section 5.1 and the passage time P; see Figure 7.1. In this network there are two classes of jobs: class 1 jobs at center 1 and class 2 jobs at center 2. Since each center sees only one job class, by taking into account the fixed number of jobs in the network, for $t\geq 0$ we can define Z(t) to be the number of jobs waiting or in service at center 1 at time t. Then the process $\underset{\sim}{X}=\{(Z(t),N(t)):t\geq 0\}$, where N(t) is the position of the marked job in the job stack at time t, has state space

$$E = \{(i,j):0\leq i\leq N, 1\leq j\leq N\} .$$

For the passage time P, the sets A_1 and A_2 defining the starts of passage times for the marked job are

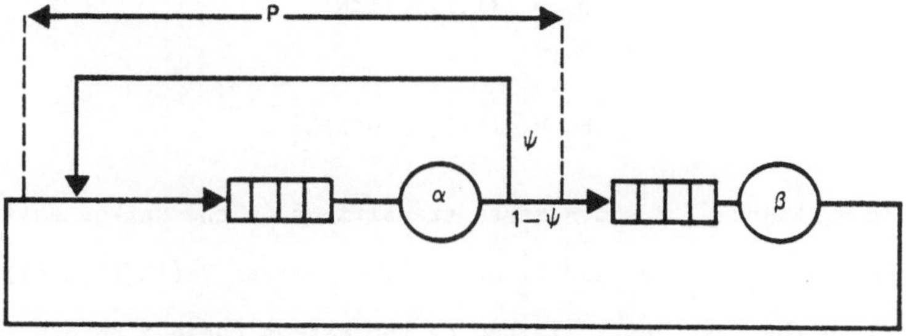

(i) α and β services are not interruptable
(ii) Routing determined by binary valued random variable ψ

Figure 7.1. Closed network of queues

$$A_1 = \{(i,N):0\le i<N\}$$

and

$$A_2 = \{(i,1):0<i\le N\} . \tag{7.2.5}$$

Similarly, the sets B_1 and B_2 defining the terminations of the passage time P are

$$B_1 = \{(i,i):0<i\le N\}$$

and

$$B_2 = \{(i-1,i):0<i\le N\} . \tag{7.2.6}$$

For N=2 jobs, Figure 7.2 shows state transitions in the Markov chain $\underset{\sim}{X}$ and the subsets A_1, A_2, B_1 and B_2 of E. The process $\underset{\sim}{V}=\{(L(t),X(t)):t\ge 0\}$, where L(t) is the last state visited by the Markov chain $\underset{\sim}{X}$ before jumping to X(t), has state space

$$F = \{(i,j,i+1,j+1):0\le i<N,\ 1\le j<N\} \cup \{(i,N,i+1,1):0\le i<N\} \cup$$
$$\{(i,j,i-1,j):0<i\le N,\ 1\le j\le N\} \cup \{(i,i,i,1):1<i\le N\} .$$

The subsets of F defining the starts and terminations of passage times for the marked job are

$$S = \{(i,N,i+1,1):0\le i<N\}$$

and

$$T = \{(i,i,i-1,i):0<i\le N\} . \tag{7.2.7}$$

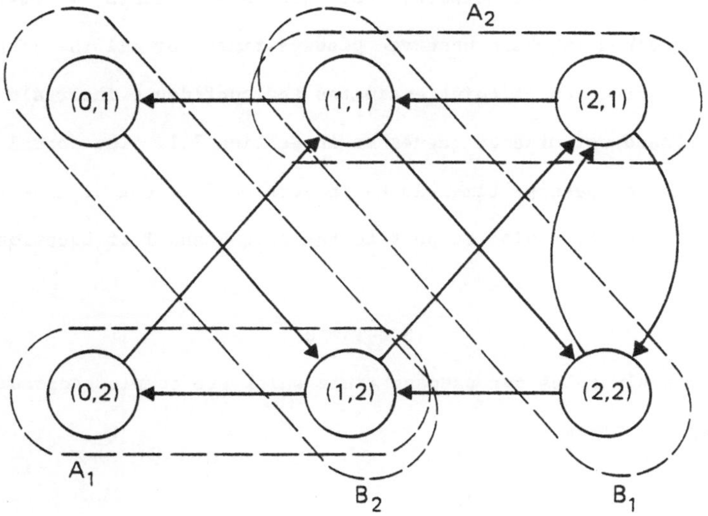

Figure 7.2. State transitions in Markov chain X and subsets of E for passage time P

8.0. THE DECOMPOSITION METHOD

In this section we concentrate on passage times through a subnetwork of a given closed network of queues, i.e., passage times which are not complete circuits in the network. For such passage times, we develop an estimation method in which observed passage times for all the jobs enter into the construction of point estimates and confidence intervals. We consider closed networks of queues as in Section 7.1. The formal specification of passage times is as in Section 7.1, but we make the additional assumption with respect to the sets S and T of Equation (7.2.3) that

$$S \cap T = \phi \ ;$$

this effectively rules out passage times which are complete circuits, i.e., response times.

The basis of the decomposition method for estimation of passage times through a subnetwork is simulation of the network in random blocks defined by the terminations of certain passage times. The distinguished passage times are those that (i) terminate when no other passage times are underway, and (ii) leave a fixed configuration of the job stack defined by Equation (7.1.1). These terminations serve to decompose the sequence of passage times for all of the jobs into independent and identically distributed blocks.

We denote by $\{P_n^0 : n \geq 1\}$ the sequence of passage times (irrespective of job identity) enumerated in order of passage time start. As before, we let f be a real-valued (measurable) function with domain R_+, and the goal of the simulation is the estimation of

$$r^0(f) = E\{f(P^0)\} ,$$

where $P_n^0 \Rightarrow P^0$. Note that by the results of Appendix 1, $P^0=P$, the limiting passage time for (any) marked job.

8.1. Simulation for Passage Times Through a Subnetwork

The main result of this section is that we can obtain point and interval estimates of the quantity $r^0(f)$ as follows.

(8.1.1) ALGORITHM.

1. Select a configuration z^0 of the job stack at which a passage time terminates and there are no other passage times underway. Begin the simulation with this configuration of the job stack at time 0.
2. Carry out the simulation for a fixed number n of blocks defined by the successive terminations of passage times <u>irrespective of job identity</u> which leave the job stack in the (fixed) configuration z^0.
3. In each block, measure the passage times for all the jobs and record these along with the number of passage times observed in the block.
4. Denote by K_m^0 the number of passage times observed in the mth block and compute $Y_m^0(f)$, the sum of the quantities $f(P_j^0)$ for the passage times P_j^0 in the mth block, e.g.,

$$Y_1^0(f) = \sum_{j=1}^{K_1^0} f(P_j^0) .$$

5. Take as a point estimate (based on n blocks) of $r^0(f)$ the quantity

$$\hat{r}_n^0(f) = \overline{Y}_n^0(f)/\overline{K}_n^0 ,$$

where

$$\overline{Y}_n^0(f) = n^{-1}\sum_{m=1}^{n} Y_m^0(f)$$

and

$$\overline{K}_n^0 = n^{-1}\sum_{m=1}^{n} K_m^0 .$$

6. Take as a $100(1-2\gamma)\%$ confidence interval (based on n blocks) for $r^0(f)$ the interval

$$\hat{I}_n(f) = \left[\hat{r}_n^0(f)-z_{1-\gamma}s_n/\left(\overline{K}_n^0 n^{1/2}\right),\ \hat{r}_n^0(f)+z_{1-\gamma}s_n/\left(\overline{K}_n^0 n^{1/2}\right)\right] .$$

Here s_n is the quantity

$$s_n = \left\{s_{11}(n)-2\hat{r}_n^0(f)s_{12}(n)+\left(\hat{r}_n^0(f)\right)^2 s_{22}(n)\right\}^{1/2} ,$$

where

$$s_{11}(n) = (n-1)^{-1}\sum_{m=1}^{n}\left(Y_m^0(f)-\overline{Y}_n^0\right)^2 ,$$

$$s_{22}(n) = (n-1)^{-1}\sum_{m=1}^{n}\left(K_m^0-\overline{K}_n^0\right)^2 ,$$

and

$$s_{12}(n) = (n-1)^{-1}\sum_{m=1}^{n}\left(Y_m^0(f)-\overline{Y}_n^0\right)\left(K_m^0-\overline{K}_n^0\right) .$$

The quantity $z_{1-\gamma}=\Phi^{-1}(1-\gamma)$, where $\Phi(\cdot)$ is the distribution function of a standardized normal random variable.

8.2. The Underlying Stochastic Structure

We begin by labelling the jobs from 1 to N, and for $i=1,2,\ldots,N$, denote by $N^i(t)$ the position of job i in the linear job stack at time t. Then, in terms of the vector $Z(t)$ as defined in Equation (2.1.1), we set

$$X^i(t) = (Z(t),N^i(t))$$

and

$$X^0(t) = (Z(t),N^1(t),N^2(t),\ldots,N^N(t)) \ . \tag{8.2.1}$$

Each of the processes $\underset{\sim}{X}^i=\{X^i(t):t\geq 0\}$ $(i=1,2,\ldots,N)$ and $\underset{\sim}{X}^0=\{X^0(t):t\geq 0\}$ is an irreducible, positive recurrent continuous time Markov chain. We denote the state space of the process $\underset{\sim}{X}^i$ [respectively $\underset{\sim}{X}^0$] by E^i [respectively E^0].

Next we let $L^i(t)$ [respectively $L^0(t)$] denote the last state visited by the Markov chain $\underset{\sim}{X}^i$ [respectively $\underset{\sim}{X}^0$] before jumping to $X^i(t)$ [respectively $X^0(t)$], and for $i=1,2,\ldots,N$ and $t\geq 0$ define

$$V^i(t) = (L^i(t),X^i(t))$$

and

$$V^0(t) = (L^0(t),X^0(t)) \ . \tag{8.2.2}$$

The process $\underset{\sim}{V}^0=\{V^0(t):t\geq 0\}$ is the fundamental stochastic process of the simulation for passage times through a subnetwork. Note that incorporation of the component $L^0(t)$ into the definition of $\underset{\sim}{V}^0$ is necessary for detection of the starts and terminations of passage times. Since each of the processes $\underset{\sim}{X}^i$ and $\underset{\sim}{X}^0$ is an irreducible, positive recurrent continuous time Markov chain, so is each of the processes $\underset{\sim}{V}^i=\{V^i(t):t\geq 0\}$ and $\underset{\sim}{V}^0$. We denote the state spaces of $\underset{\sim}{V}^i$ and $\underset{\sim}{V}^0$ by F^i and F^0, respectively. As in Equation (7.2.3), we define subsets S^i and T^i of F^i according to

$$S^i = \{(k,m)\in F^i : k\in A_1,\ m\in A_2\}$$

and

$$T^i = \{(k,m)\in F^i : k\in B_1,\ m\in B_2\}\ . \tag{8.2.3}$$

Thus, the entrances of $\underset{\sim}{V}^i$ to S^i [respectively T^i] correspond to the starts [respectively terminations] of passage times for job i. We also define two subsets S^0 and T^0 of F^0 according to

$$S^0 = \{(z,n_1,\ldots,n_N,z',n_1',\ldots,n_N')\in F^0:\ \text{for some } k,\ (z,n_k)\in A_1 \text{ and } (z',n_k')\in A_2\}$$

and

$$T^0 = \{(z,n_1,\ldots,n_N,z',n_1',\ldots,n_N')\in F^0:\ \text{for some } k,\ (z,n_k)\in B_1 \text{ and } (z',n_k')\in B_2\}\ . \tag{8.2.4}$$

The entrances of $\underset{\sim}{V}^0$ to the set S^0 correspond to the starts of passage times (irrespective of job identity) and the entrances of $\underset{\sim}{V}^0$ to the set T^0 correspond to the terminations. Thus, from a simulation of the process $\underset{\sim}{V}^0$, it is possible to measure the passage times for each of the jobs.

Now consider $\{P_n^0 : n\geq 1\}$ the sequence of passage times (irrespective of job identity), enumerated in order of passage time start. Formal definition of the sequence $\{P_n^0\}$ is in terms of sequences of starts and terminations of passage times for each of the jobs; the definitions of the latter involve entrances of $\underset{\sim}{V}^i$ to the sets S^i and T^i of Equation (8.2.3) and are analogous to Equation (7.1.3). Recall that the goal of the simulation is

$$r^0(f) = E\{f(P^0)\} , \tag{8.2.5}$$

where $P_n^0 \Rightarrow P^0$, and f is a real-valued measurable function with domain R_+.

We carry out the simulation in random <u>blocks</u> of the process $\underset{\sim}{V}^0$ defined by the successive entrances of $\underset{\sim}{V}^0$ to a fixed set of states U^0. Entrances of $\underset{\sim}{V}^0$ to the set U^0 correspond to the terminations of passage times (irrespective of job identity) which occur when no other passage times are underway, and which leave a fixed configuration of the job stack.

Formally, let D be the state space of the process $\underset{\sim}{Z}=\{Z(t):t\geq 0\}$ defined by Equation (3.2.1), and denote by C the set of (center, class) pairs in the network. We define a function h taking values in C and having domain $D\times\{1,2,\ldots,N\}$ as follows. For $z\in D$ and $n\in\{1,2,\ldots,N\}$, the value of $h(z,n)$ is (i,j) when the job in position n of (the job stack) z is of class j at center i. Now consider the embedded jump chain $\{V_k:k\geq 0\}$ associated with the continuous time Markov chain $\underset{\sim}{V}$ of Equation (7.2.2). For states $v',v''\in F$ the state space of $\{V_k:k\geq 0\}$, we write $v' \rightsquigarrow v''$ when v'' is accessible from v', i.e., when for some $n\geq 1$, the probability starting from v' of entering v'' on the nth step is positive. Similarly, for any subset L of F we write $v' \overset{L}{\rightsquigarrow} v''$ when v'' is accessible from v' under the taboo L; this means (cf. Chung (1967), pp.45, 48) that for some $n\geq 1$, there is a positive probability, starting from state v', of entering state v'' on the nth step under the restriction that none of the states in L is entered in between (exclusive of both ends).

Next, we define a subset H of C according to

$$H = \{(i,j)\epsilon C: \text{ for some } (z,n,z',n')\epsilon T-S,\ h(z',n') = (i,j)\} \cup$$
$$\{(i,j)\epsilon C: \text{ for some } (z,n,z',n')\epsilon F-(T\cup S),\ v'\epsilon T \text{ and } v''\epsilon S,$$
$$v' \overset{S}{\sim} (z,n,z',n'),\ (z,n,z',n') \overset{S}{\sim} v'' \text{ and } h(z',n') = (i,j)\}$$

Thus, the set H is $H_1 \cup H_2$, where a (center, class) pair is in H_1 [respectively H_2] if it is possible for the marked job to be of class j at center i when the passage time specified by the sets A_1, A_2, B_1, and B_2 terminates [respectively is not underway]. Note that the set H is nonempty since by assumption $S\cap T=\phi$ and thus H_1 is nonempty.

Now define a subset D^0 of D, the state space of $\{Z(t):t\geq 0\}$, according to

$$D^0 = \{z\epsilon D: h(z,n)\epsilon H \text{ for } n=1,2,\ldots,N \text{ and for some } n,\ h(z,n)\epsilon H_1\} .$$

Elements of the set D^0 correspond to configurations of the job stack upon termination of a passage time with no other passage times underway. The set D^0 is nonempty since H is nonempty. Therefore, we can select an element z^0 of D^0, and in terms of this fixed z^0, finally define the set U^0 according to

$$U^0 = \{(z,n_1,\ldots,n_N,z',n_1',\ldots,n_N')\epsilon T^0: z'=z^0\} , \tag{8.2.6}$$

where T^0 is given by Equation (8.2.4).

For convenience, we assume that $V^0(0)\epsilon U^0$. The random times $\{\gamma_m^0:m\geq 1\}$ denote the lengths of the successive blocks (returns to the set U^0) for $\{V_n^0:n\geq 0\}$ the embedded jump chain associated with $\underset{\sim}{V}^0$, and we define $\delta_0^0=0$ and $\delta_m^0=\gamma_1^0+\ldots+\gamma_m^0$, $m\geq 1$.

The number of passage times K_1^0 in the first block of the process $\underset{\sim}{V}^0$ is

$$K_1^0 = \sum_{j=0}^{\delta_1^0 - 1} 1_{\{V_j^0 \in S^0\}} ,$$

and we denote the analogous quantity in the mth block of $\underset{\sim}{V}^0$ by K_m^0. Note that within each block of $\underset{\sim}{V}^0$ defined by the entrances to the set U^0, at least one passage time starts and terminates.

Next, we let $Y_m^0(f)$ be the sum of the quantities $f(P_j^0)$ over the passage times P_j^0 in the mth block of $\underset{\sim}{V}^0$, for example,

$$Y_1^0(f) = \sum_{j=1}^{K_1^0} f(P_j^0) .$$

Proposition (8.2.7) contains the key observation.

(8.2.7) PROPOSITION. The sequence of pairs of random variables $\{(Y_m^0(f), K_m^0) : m \geq 1\}$ are independent and identically distributed.

The argument used in Appendix 1 shows that $P_n^0 \Rightarrow P^0$ as $n \to \infty$, and that this random variable P^0 is the same random variable as the limiting passage time P of (any) marked job. For the function f appearing in the definition of $r^0(f)$, let D(f) denote the set of discontinuities of f. Assuming that $P\{P^0 \in D(f)\} = 0$ and using Lemma (2.1.9), it follows that as $n \to \infty$, $f(P_n^0) \Rightarrow f(P^0)$. Finally, standard arguments (cf. Appendix 2) yield a ratio formula for $r^0(f)$.

(8.2.8) PROPOSITION. Assume that $E\{|f(P^0)|\}<\infty$. Then

$$r^0(f) = E\{f(P^0)\} = E\{Y_1^0(f)\}/E\{K_1^0\} .$$

With the ratio formula of Proposition (8.2.8) and the result (Proposition (8.2.7)) that the pairs of random variables $\{(Y_m^0(f),K_m^0):m\geq 1\}$ are i.i.d., the regenerative method applies; from a fixed number of blocks we obtain the point estimate $r_n^0(f)=\overline{Y}_n^0(f)/\overline{K}_n^0$ and an associated confidence interval for $r^0(f)$.

(8.2.9) EXAMPLE. Consider the model of Section 5.1 and Example 7.2.4. First observe that for the passage time P, the sets S and T (given by Equation (7.2.7)) are disjoint. The process $\underset{\sim}{X}^0=\{(Z(t),N^1(t),\ldots,N^N(t)):t\geq 0\}$ has state space E^0, where

$$E^0 = \{(i,n_1,\ldots,n_N):0\leq i\leq N;\ 1\leq n_1,\ldots,n_N\leq N;\ n_i\neq n_j \text{ for } i\neq j\} .$$

The underlying continuous time process $\underset{\sim}{V}^0=\{(L^0(t),X^0(t)):t\geq 0\}$, where $L^0(t)$ is the last state visited by the Markov chain $\underset{\sim}{X}^0$ before jumping to $X^0(t)$, has state space F^0. The subsets of F^0 defining starts and terminations of passage times are

$$S^0 = \{(i,n_1,\ldots,n_N,i+1,n_1',\ldots,n_N'):\ \text{for exactly one } j^*,\ n_{j^*}=N \text{ and } n'_{j^*}=1;\ 1\leq n_j<N \text{ and } n_j'=n_j+1,\ j\neq j^*;\ n_i\neq n_j \text{ for } i\neq j\}$$

and

$$T^0 = \{(i,n_1,\ldots,n_N,i-1,n_1',\ldots,n_N'):0<i\leq N;\ 1\leq n_j=n_j'\leq N, \text{ for } 1\leq j\leq N;\ n_i\neq n_j \text{ for } i\neq j\} .$$

For this network the process $\underset{\sim}{Z}=\{Z(t):t\geq 0\}$ has state space $D=\{0,1,\ldots,N\}$. The set C of (center, class) pairs is $\{(1,1),(2,2)\}$, the set $H_2=\phi$ and $H=H_1=\{(2,2)\}$. The set $D^0=\{0\}$, and since D^0 is a singleton, we must select $z^0=0$. For $z^0=0$, the set U^0 defining the blocks of the process $\underset{\sim}{V}^0$ is

$$U^0 = \{(1,n_1,\ldots,n_N,0,n_1,\ldots,n_N):1\leq n_1,\ldots,n_N\leq N;\ n_i\neq n_j;\ \text{for } i\neq j\} .$$

Now consider the case of $N=2$ jobs. The state space F^0 of the continuous time Markov chain $\underset{\sim}{V}^0$ has ten elements, and the set

$$U^0 = \{(1,1,2,0,1,2),\ (1,2,1,0,2,1)\} ;$$

see Figure 8.1. A portion of a sample path for $\underset{\sim}{V}^0$ appears in Figure 8.2, and Figure 8.3 shows the corresponding decomposition of the sequence $\{P_n^0\}$.

For passage times through a subnetwork, the decomposition method provides an alternative to marked job simulation. Since observed passage times for all of the jobs enter in the construction of these point and interval estimates, we would expect this method to have greater statistical efficiency than the marked job method. In this connection, the calculation of theoretical values for variance constants entering into central limit theorems used to obtain confidence intervals from passage time simulations is of interest. These calculations are the subject of Section 9.

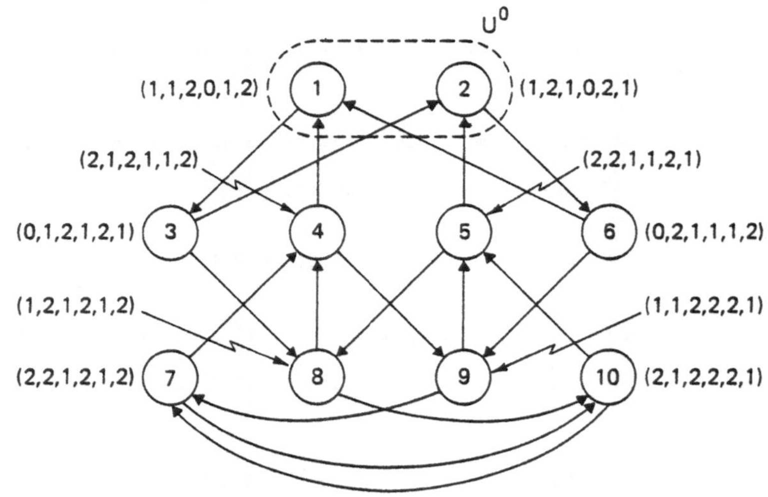

Figure 8.1. State transitions in Markov chain $V^0$and subset U^0 of F^0

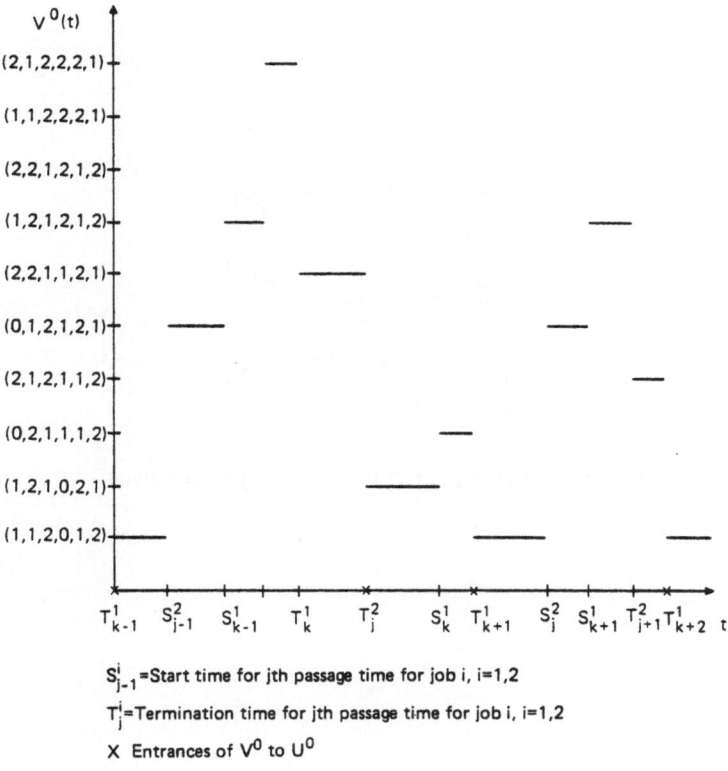

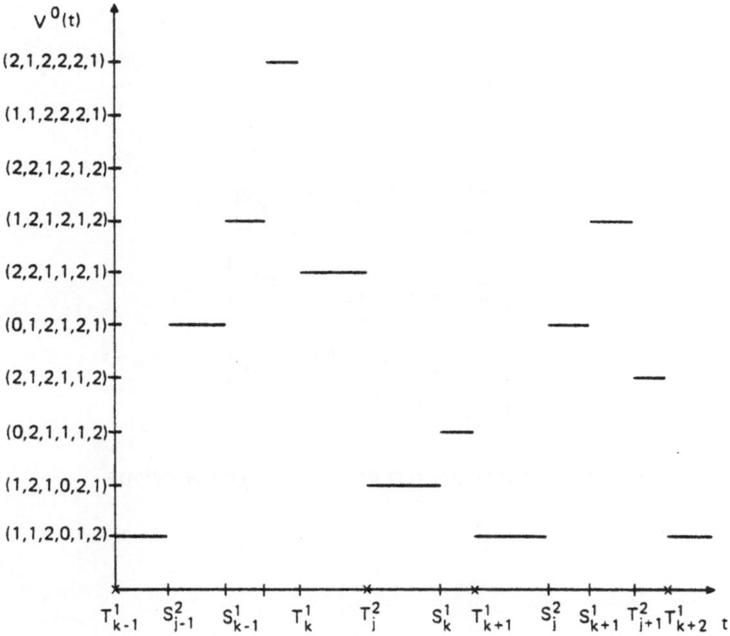

S^i_{j-1}=Start time for jth passage time for job i, i=1,2

T^i_j=Termination time for jth passage time for job i, i=1,2

X Entrances of V^0 to U^0

Figure 8.2. Sample path for process V^0

$$K_m^0 = 2 \quad K_{m+1}^0 = 1 \quad K_{m+2}^0 = 2$$

$$\cdots \quad \left| P_n^0 \quad P_{n+1}^0 \right| P_{n+2}^0 \left| P_{n+3}^0 \quad P_{n+4}^0 \right| \quad \cdots$$

$$\cdots \quad \left| P_j^2 \quad P_k^1 \right| P_{k+1}^1 \left| P_{j+1}^2 \quad P_{k+2}^1 \right| \quad \cdots$$

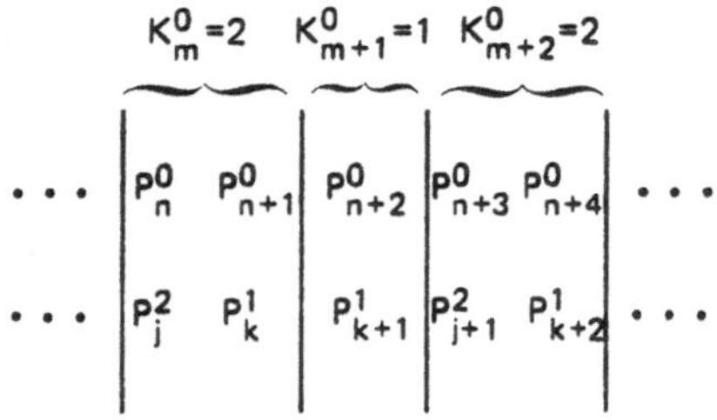

Figure 8.3. Decomposition of sequence of passage times

9.0. EFFICIENCY OF SIMULATION

We consider in this section the calculation of theoretical values for variance constants entering into the central limit theorems used in previous sections to obtain confidence intervals for passage time characteristics. Using results of Hordijk, Iglehart and Schassberger (1976) for the calculation of moments in discrete time and continuous time Markov chains, we compute theoretical values for mean passage times. We do this first for the marked job method (in the stochastic setting of Section 7), and then for the decomposition method of Section 8. For passage times where both estimation methods apply, these results provide a firm basis for a comparison of statistical efficiency. The calculations also make it possible to assess the efficacy of the marked job method for simulation of response times.

9.1. Theoretical Values for Finite State Markov Chains

Following Hordijk, Iglehart, and Schassberger (1976), we first consider discrete time Markov chains, and let $\{X_k:k\geq 0\}$ be an irreducible Markov chain with finite state space $E=\{0,1,\ldots,N\}$ and one-step transition matrix

$$\underline{P} = \{p_{ij}:i,j\in E\} .$$

For this chain, we let p_{ij}^n denote the n-step transition probability from state i to state j, and recall that for $n\geq 1$,

$$\underline{P}^n = \{p_{ij}^n:i,j\in E\} .$$

Throughout this section we use the following notation. For a fixed state $i\in E$, $P_i\{\cdot\}$ denotes the conditional probability associated with starting the chain in state i, and $E_i\{\cdot\}$ denotes the corresponding conditional expectation. For $j\in E$, the state space of $\{X_k:k\geq 0\}$, and $n\geq 1$ we let $\beta_n(j)$ denote the nth entrance time of $\{X_k:k\geq 0\}$ to state j, e.g.,

$$\beta_1(j) = \min\{k\geq 1:X_k=j\}$$

and let $\alpha_1(j)=\beta_1(j)$ and $\alpha_n(j)=\beta_n(j)-\beta_{n-1}(j)$, $n>1$. This notation is consistent with that introduced in Section 2 for regenerative processes. Note that $\{\beta_n(j):n\geq 1\}$ is a (possibly delayed) renewal process since a finite state, irreducible Markov chain is necessarily positive recurrent and therefore returns to every state $j\in E$ infinitely often with probability one. If $X_0=j$, the process $\{\beta_n(j):n\geq 1\}$ is an ordinary renewal process.

We consider vectors such as $(v(0),v(1),\ldots,v(N))$ to be column vectors. Real-valued functions, such as f and g, having domain E are viewed in this way and denoted by $\underline{f}$ and $\underline{g}$. In this context the symbol $E\{\cdot\}$ denotes the vector

$$(E_0\{\cdot\},E_1\{\cdot\},\ldots,E_N\{\cdot\}) .$$

In addition (for vectors u and v) the symbol $u\circ v$ denotes the vector

$$(u(0)v(0),u(1)v(1),\ldots,u(N)v(N)) .$$

For a matrix $\underline{A}=(a_0,a_1,\ldots,a_m)$, we let

$$u\circ\underline{A} = \underline{A}\circ u = (u\circ a_0,u\circ a_1,\ldots,u\circ a_m) .$$

Finally, for a matrix $\underline{B}=(b_0,b_1,\ldots,b_m)$, we let

$$\underline{A}\circ\underline{B} = (a_0\circ b_0,a_1\circ b_1,\ldots,a_m\circ b_m) .$$

We first consider computation of the values of $E_i\{Y_1(f)\}$ and $E_i\{Y_1(f)Y_1(g)\}$ for general real-valued functions f and g having domain E, and $i\epsilon E$. Accomplishing this (and similar computations for continuous time Markov chains) makes it possible to obtain theoretical values for mean passage times; in addition we can obtain values for the variance constants which enter into central limit theorems used to obtain confidence intervals in simulations by the marked job and decomposition methods.

For the discrete time Markov chain $\{X_k:k\geq 0\}$, we consider here only cycles of the regenerative process formed by the successive entrances to state 0, and henceforth suppress the 0 in the notation $\beta_n(0)$, $\alpha_n(0)$, etc. Note that this is no real restriction, and that equally well we could choose any other state $i\epsilon E$. For $i,j\epsilon E$ and $n=0,1,\ldots$, let

$${}_0p_{ij}^n = P_i\{\alpha_1>n,\ X_n=j\} ,$$

and set

$${}_0\underline{P}^n = \{{}_0p_{ij}^n : i,j\epsilon E\} .$$

We obtain

$${}_0\underline{P}^1 = {}_0\underline{P}$$

from $\underline{P}$ by setting the 0-column of $\underline{P}$ equal to 0. It is easy to see that ${}_0\underline{P}^n$ is the matrix product of n copies of ${}_0\underline{P}$, and that for all $n\geq 1$,

$${}_0p_{i0}^n = 0 .$$

Theorem (9.1.1) is due to Hordijk, Iglehart, and Schassberger (1976). For any real-valued function f with domain E, we define

$$Y_1(f) = \sum_{k=0}^{\beta_1 - 1} f(X_k) .$$

(9.1.1) THEOREM. For an irreducible, finite state discrete time Markov chain with transition matrix $\underline{P}$,

$$E\{Y_1(f)\} = (\underline{I} - {}_0\underline{P})^{-1}\underline{f} \tag{9.1.2}$$

and

$$E\{Y_1(f)Y_1(g)\} = (\underline{I} - {}_0\underline{P})^{-1}\underline{h} , \tag{9.1.3}$$

where $\underline{h} = \underline{f} \circ E\{Y_1(g)\} + \underline{g} \circ E\{Y_1(f)\} - \underline{f} \circ \underline{g}$.

Now we consider continuous time Markov chains and let $\underset{\sim}{X} = \{X(t) : t \geq 0\}$ be a Markov chain with finite state space $E = \{0, 1, \ldots, N\}$, transition matrix $\underline{P}(t) = \{p_{ij}(t) : i, j \in E\}$ and matrix of infinitesimal transition parameters $\underline{Q} = \{q_{ij} : i, j \in E\}$. Recall that in a continuous time Markov chain the matrix $\underline{Q} = \underline{P}'(0)$ is the given data. In general, $\underline{P}(t)$ is difficult to calculate and is rarely available explicitly. The exponentially distributed holding time in any state $i \in E$ has rate parameter $q_i = -q_{ii}$. For all $i \in E$, we assume that $0 < q_i < \infty$, i.e., that all states are stable and nonabsorbing, and in addition that

$$\sum_{j=0}^{N} q_{ij} = 0 .$$

This last assumption guarantees that, starting from any state $i \in E$, the Markov chain $\underset{\sim}{X}$ makes a transition to a next state $j \in E$. We now form the jump matrix $\underline{R}=\{r_{ij}\}$ of the chain, defining the elements r_{ij} according to

$$r_{ij} = \begin{cases} q_{ij}/q_i \ , & j \neq i \\ 0 \ , & j = i \end{cases} .$$

We assume that the jump matrix $\underline{R}$ is irreducible (and therefore positive recurrent). This is equivalent to the continuous time Markov chain $\underset{\sim}{X}$ being irreducible. As in the case of a discrete time Markov chain, we let $P_i\{\cdot\}$ and $E_i\{\cdot\}$ denote the conditional probability and conditional expectation associated with starting in state $i \in E$. For $j \in E$ and $n \geq 1$, we let $\beta_n(j)$ denote the nth entrance time of $\underset{\sim}{X}$ to state j, i.e.,

$$\beta_1(j) = \inf\{s>0 : X(s-) \neq j,\ X(s)=j\} .$$

We now consider the computation of $E_i\{Y_1(f)\}$ and $E_i\{Y_1(f)Y_1(g)\}$ for real valued functions f and g with domain E. As in the case of discrete time Markov chains, we restrict attention to regenerative cycles formed by the successive entrances to state 0, and suppress the 0 in our notation. For $t \geq 0$, we let

$${}_0p_{ij}(t) = P_i\{\alpha_1 > t,\ X(t)=j\} ,$$

$${}_0\underline{P}(t) = \{{}_0p_{ij}(t) : i,j \in E\} ,$$

and, for $n \geq 0$, construct the matrix ${}_0\underline{R}^n$ from $\underline{R}$ in the same manner as we constructed ${}_0\underline{P}^n$ from $\underline{P}$ in the discrete time case.

For a real-valued function f defined on E, we define $Y_1(f)$ according to

$$Y_1(f) = \int_0^{\beta_1} f(X(t))dt ,$$

and let $\underline{q}^{-1}$ be the column vector

$$\underline{q}^{-1} = (q_0^{-1}, q_1^{-1}, \ldots, q_N^{-1}) .$$

Theorem (9.1.4) is due to Hordijk, Iglehart, and Schassberger (1976).

(9.1.4) THEOREM. For an irreducible, finite state continuous time Markov chain with jump matrix $\underline{R}$ and vector q of rate parameters for holding times,

$$E\{Y_1(f)\} = E\left\{\int_0^\infty {}_0\underline{P}(t)\underline{f}\, dt\right\} = (\underline{I}-{}_0\underline{R})^{-1}(\underline{f}\circ\underline{q}^{-1}) \tag{9.1.5}$$

and

$$E\{Y_1(f)Y_1(g)\} = E\left\{\int_0^\infty {}_0\underline{P}(t)\underline{h}\, dt\right\} = (\underline{I}-{}_0\underline{R})^{-1}(\underline{h}\circ\underline{q}^{-1}) \tag{9.1.6}$$

where $\underline{h}=\underline{f}\circ E\{Y_1(g)\}+\underline{g}\circ E\{Y_1(f)\}$.

We now show how to use the results of Theorems (9.1.1) and (9.1.4) to assess the statistical efficiency of simulation by the marked job method for mean passage times.

9.2. Variance Constants for the Marked Job Method

We consider closed networks of queues and passage times as in Section 7.1. For $t\geq 0$, the state vector of the network is

$$X(t) = (Z(t),N(t)) , \qquad (9.2.1)$$

where Z(t) (of Equation (7.1.1)) corresponds to the linear job stack, and N(t) is the position in the job stack of the marked job at time t. Recall that the process $\underset{\sim}{X}=\{X(t):t\geq 0\}$ is an irreducible, positive recurrent Markov chain with state space E. As before, we denote by L(t) the last state visited by the Markov chain $\underset{\sim}{X}$ before jumping to X(t), and the process $\underset{\sim}{V}=\{V(t):t\geq 0\}$ defined by

$$V(t) = (L(t),X(t)) \qquad (9.2.2)$$

is the fundamental stochastic process of the passage time simulation.

Recall that the process $\underset{\sim}{V}$ has a state space F consisting of all pairs of states (i,j), $i,j\in E$ for which a transition in $\underset{\sim}{X}$ from state i to state j can occur with positive probability. Since $\underset{\sim}{X}$ is an irreducible, positive recurrent Markov chain, so is $\underset{\sim}{V}$, and the entrances of $\underset{\sim}{V}$ to the fixed subset S [respectively T] (defined by Equation (7.2.3)) of the state space F correspond to the starts [respectively terminations] of passage times for the marked job.

As in Section 7.2, we select a (fixed) state of S, designated state 0, and assume that V(0)=0. To estimate the quantity r(f) of Equation (7.2.1), the marked job method prescribes that we carry out the simulation of $\underset{\sim}{V}$ in 0-cycles defined by the successive returns to state 0; within each cycle we record the number of passage times of the marked job and measure each of these passage times.

The key results of Section 7.2 leading to point estimates and confidence intervals for the quantity r(f) are that the pairs of random variables

$$\{(Y_k(f),M_k):k\geq 1\} \tag{9.2.3}$$

are independent and identically distributed, and that

$$r(f) = E_0\{Y_1(f)\}/E_0\{M_1\} . \tag{9.2.4}$$

Recall that M_k is the number of passage times for the marked job in the kth 0-cycle and $Y_k(f)$ is the sum of the values of the function f for the passage times of the marked job in this cycle.

Given Equations (9.2.3) and (9.2.4), the regenerative method provides (from a fixed number n of 0-cycles) the point estimate

$$\hat{r}_n(f) = \overline{Y}_n(f)/\overline{M}_n .$$

The associated confidence interval for r(f) follows from the central limit theorem

$$\frac{n^{1/2}\{\hat{r}_n(f)-r(f)\}}{\sigma(f)/E_0\{M_1\}} \Rightarrow N(0,1) , \tag{9.2.5}$$

where

$$\sigma^2(f) = \mathrm{var}\{Y_1(f)-r(f)M_1\} . \tag{9.2.6}$$

For calculation of theoretical values, we restrict attention to the mean passage time; thus, the function f in the definition of r(f) is the identity function. Using the results of Section 9.1, we show how to compute the value of the mean passage time r and the corresponding variance constant σ^2 appearing in the central limit theorem of Equation (9.2.5). These computations rest on the definition of two particular functions (denoted f^* and g^*) having domain F and taking values in the set $\{0,1\}$.

We define the function f^* to be the indicator function, 1_S, of the set S which defines the starts of passage times for the marked job; i.e., for $(z,n,z',n')\in F$,

$$f^*(z,n,z',n') = 1_S(z,n,z',n) \ . \tag{9.2.7}$$

Proposition (9.2.8) follows directly from Theorem (9.1.1).

(9.2.8) PROPOSITION. Let f^* be the function defined by Equation (9.2.7), and $\underline{R}$ the transition matrix of the discrete time Markov chain $\{V_k:k\geq 0\}$. Then

$$E\{Y_1(f^*)\} = E\left\{\sum_{k=0}^{\delta_1-1} f^*(V_k)\right\} = (\underline{I}-{}_0\underline{R})^{-1}\underline{f}^*$$

and

$$E\{(Y_1(f^*))^2\} = (\underline{I}-{}_0\underline{R})^{-1}\underline{h}^* \ ,$$

where δ_1 is the time of the first return to the state 0 in $\{V_k:k\geq 0\}$ and $\underline{h}^*=2\underline{f}^*\circ E\{Y_1(f^*)\}-\underline{f}^*\circ\underline{f}^*$.

We use Proposition (9.2.8) and the definition of M_1 to obtain the quantities $E_0\{M_1\}$ and $E_0\{M_1^2\}$ according to

$$E_0\{M_1\} = E_0\{Y_1(f^*)\} \tag{9.2.9}$$

and

$$E_0\{M_1^2\} = E_0\{(Y_1(f^*))^2\} . \tag{9.2.10}$$

For an element $(z,n,z',n')\epsilon F$, the value of the function g^* is 1 if a passage time for the marked job starts or is underway when the configuration of the job stack is z' and the marked job is in position n'; the value of g^* is 0 otherwise. Formally, let D be the state space of the process $\underset{\sim}{Z}=\{Z(t):t\geq 0\}$ appearing in Equation (9.2.1). As in Section 8, for $z\epsilon D$ and $n\epsilon\{1,2,\ldots,N\}$, we write $h(z,n)=(i,j)$ when the job in position n in the job stack z is of class j at center i. Now consider the embedded jump chain $\{V_k:k\geq 0\}$ associated with the continous time Markov chain $\underset{\sim}{V}$ of Equation (9.2.2). For states $v',v''\epsilon F$, the state space of $\{V_k:k\geq 0\}$, we write $v' \sim v''$ when v'' is accessible from v', i.e., when for some $n\geq 1$, the probability starting from v' of entering v'' on the nth step is positive. Similarly, for any subset I of F we write $v' \overset{I}{\sim} v''$ when v'' is accessible from v' under the taboo I.

Denoting the set of (center, class) pairs in the network by C, we define a subset G of C according to

$$\begin{aligned} G = \{(i,j)\epsilon C:\ &\text{for some } (z,n,z',n')\epsilon S,\ h(z',n')=(i,j)\} \cup \\ \{(i,j)\epsilon C:\ &\text{for some } (z,n,z',n')\epsilon F-(S\cup T),\ v'\epsilon S \text{ and } v''\epsilon T\ , \\ &v' \overset{T}{\sim} (z,n,z',n'),\ (z,n,z',n') \overset{T}{\sim} v'' \text{ and } h(z',n')=(i,j)\} . \end{aligned}$$

Thus, the set G is $G_1 \cup G_2$, where a (center, class) pair is in the set G_1 [respectively G_2] if it is possible for the marked job to be of class j at center i when the passage time specified by the sets A_1, A_2, B_1, and B_2 starts [respectively is underway].

Now, for $(z,n,z',n') \in F$, we define the function g^* as

$$g^*(z,n,z',n') = 1_G(h(z',n')) \ . \qquad (9.2.11)$$

(9.2.12) PROPOSITION. Let g^* be the function defined by Equation (9.2.11), and $\underline{R}$ be the jump matrix and $\underline{q}$ the vector of rate parameters for holding times in the continuous time Markov chain $\underset{\sim}{V}$. Then

$$E\{Y_1(g^*)\} = E\left\{\int_0^{\beta_1} g^*(V(s))ds\right\} = (\underline{I}-{}_0\underline{R})^{-1}(\underline{g}^* \circ \underline{q}^{-1}) \ ,$$

and

$$E\{(Y_1(g^*))^2\} = (\underline{I}-{}_0\underline{R})^{-1}(\underline{h}^* \circ \underline{q}^{-1}) \ ,$$

where β_1 is the time of the first return to the state 0 in $\underset{\sim}{V}$, and $\underline{h}^* = 2\underline{g}^* \circ E\{Y_1(g^*)\}$.

Proposition (9.2.12) follows directly from Theorem (9.1.4). We use this result together with the observation that

$$\int_0^{\beta_1} g^*(V(s))ds = \sum_{j=1}^{M_1} P_j \ , \qquad (9.2.13)$$

to obtain the quantities

$$E_0\left\{\sum_{j=1}^{M_1} P_j\right\} = E_0\{Y_1(g^*)\} \tag{9.2.14}$$

and

$$E_0\left\{\left(\sum_{j=1}^{M_1} P_j\right)^2\right\} = E_0\{(Y_1(g^*))^2\} . \tag{9.2.15}$$

Using the ratio formula, Equations (9.2.9) and (9.2.14) yield the quantity r. To obtain the variance constant σ^2 appearing in the central limit theorem (Equation (9.2.5)) for the marked job method, we require one more result.

(9.2.16) PROPOSITION. Let $\underline{R}$ be the jump matrix and $\underline{q}$ the vector of rate parameters for holding times in the continuous time Markov chain $\underset{\sim}{V}$. For the functions f^* and g^* defined by Equations (9.2.7) and (9.2.10),

$$E\left\{\int_0^{\beta_1} g^*(V(s))ds \sum_{j=0}^{\delta_1-1} f^*(V_k)\right\} = (\underline{I}-{}_0\underline{R})^{-1}\underline{h}^* ,$$

where $\underline{h}^*=[(\underline{I}-{}_0\underline{R})^{-1}\underline{f}^*]\circ(\underline{g}^*\circ\underline{q}^{-1})+[(\underline{I}-{}_0\underline{R})^{-1}(\underline{g}^*\circ\underline{q}^{-1})]\circ\underline{f}^*-(\underline{g}^*\circ\underline{q}^{-1}\circ\underline{f}^*)$.

Proposition (9.2.16) does not follow directly from Theorem (9.1.1) for discrete time Markov chains or from Theorem (9.1.4) for continuous time Markov chains, but is established by similar methods; see Iglehart and Shedler (1979b), Appendix. From Proposition (9.2.16) we obtain

$$E_0\left\{\left(\sum_{j=1}^{M_1} P_j\right)M_1\right\} = E_0\left\{\int_0^{\beta_1} g^*(V(s))ds \sum_{k=0}^{\delta_1-1} f^*(V_k)\right\} . \tag{9.2.17}$$

Then an expression for the variance constant σ^2 is

$$\sigma^2 = E_0\left\{\left(\sum_{j=1}^{M_1} P_j\right)^2\right\} - 2rE_0\left\{\left(\sum_{j=1}^{M_1} P_j\right)M_1\right\} + r^2E_0\{M_1^2\} .$$

This follows from Equations (9.2.10), (9.2.15), and (9.2.17).

When comparing the statistical efficiency of the marked job and decomposition methods, it is convenient to have a central limit theorem comparable to Equation (9.2.5) but in terms of simulation time, t, rather than number of cycles, n. Let m(t) be the number of passage times completed by time t, i.e., in the interval (0,t]. If we denote by n(t) the number of 0-cycles completed by time t, then from renewal theory, as $t\to\infty$,

$$\frac{n(t)}{t} \to \frac{1}{E_0\{\alpha_1\}}$$

with probability one, where $E_0\{\alpha_1\}$ is the expected length of a 0-cycle in $\underset{\sim}{V}$. This implies that for large t, the number of 0-cycles completed by time t is approximately $t/E_0\{\alpha_1\}$. Combining this result with Equation (9.2.5), it follows that as $t\to\infty$,

$$\frac{t^{1/2}[\{m(t)\}^{-1}\left(\sum_{i=1}^{m(t)} f(P_i)\right)-r(f)]}{(E_0\{\alpha_1\})^{1/2}\sigma(f)/E_0\{M_1\}} \Rightarrow N(0,1) .$$

This result is independent of the initial state V(0). Since the numerator in this central limit theorem is independent of the state 0 selected to form cycles, so is the denominator. Thus for the mean passage time

$$e = (E_0\{\alpha_1\})^{1/2}\sigma/E_0\{M_1\} \tag{9.2.18}$$

is the appropriate measure of statistical efficiency for the marked job method and is independent of the state $0\epsilon S$ selected to form cycles. Note that we obtain the quantity $E_0\{\alpha_1\}$ according to

$$E_0\{\alpha_1\} = E_0\{Y_1(1)\} ,$$

where 1 is the f function identically equal to one and

$$E\{Y_1(1)\} = E\left\{\int_0^{\beta_1} 1(V(s))ds\right\} = (\underline{I}-{}_0\underline{R})^{-1}(\underline{1}\circ\underline{q}^{-1}) . \tag{9.2.19}$$

9.3. Variance Constants for the Decomposition Method

We now turn to the decomposition method. As in Section 8.2, we label the jobs from 1 to N, and for $i=1,2,\ldots,N$, denote by $N^i(t)$ the position of job i in the linear job stack at time t. Then, in terms of the vector $Z(t)$ corresponding to the job stack, for $t\geq 0$ we set

$$X^i(t) = (Z(t),N^i(t))$$

and

$$X^0(t) = (Z(t),N^1(t),N^2(t),\ldots,N^N(t)) . \tag{9.3.1}$$

Recall that each of the processes $\underset{\sim}{X}^i=\{X^i(t):t\geq 0\}$ $(i=1,2,\ldots,N)$ and $\underset{\sim}{X}^0=\{X^0(t):t\geq 0\}$ is an irreducible, positive recurrent continuous time Markov chain. We denote the state space of the process $\underset{\sim}{X}^i$ [respectively $\underset{\sim}{X}^0$] by E^i [respectively E^0].

Next we let $L^i(t)$ [respectively $L^0(t)$] denote the last state visited by the Markov chain $\underset{\sim}{X}$ [respectively $\underset{\sim}{X}^0$] before jumping to $X^i(t)$ [respectively $X^0(t)$], and for $t\geq 0$ and $i=1,2,\ldots,N$, define

$$V^i(t) = (L^i(t),X^i(t))$$

and

$$V^0(t) = (L^0(t),X^0(t)) \ . \qquad (9.3.2)$$

The process $\underset{\sim}{V}^0=\{V^0(t):t\geq 0\}$ is the fundamental stochastic process of the passage time simulation.

Since each of the processes $\underset{\sim}{X}^i$ and $\underset{\sim}{X}^0$ is an irreducible, positive recurrent continuous time Markov chain, so is each of the processes $\underset{\sim}{V}^i=\{V^i(t):t\geq 0\}$ and $\underset{\sim}{V}^0$. We denote the state spaces of $\underset{\sim}{V}^i$ and $\underset{\sim}{V}^0$ by F^i and F^0, respectively. The entrances of $\underset{\sim}{V}^i$ to the fixed subset S^i [respectively T^i] of F^i (defined by Equation (8.2.3)) correspond to the starts [respectively terminations] of passage times for job i. Similarly, the successive entrances of $\underset{\sim}{V}^0$ to the fixed subset S^0 (defined by Equation (8.2.4)) of F^0 correspond to the starts of passage times irrespective of job identity, and the entrances of $\underset{\sim}{V}^0$ to the subset T^0 correspond to the terminations.

The decomposition method applies to passage times for which the sets S and T (which define the starts and terminations of passage times of a particular job) are disjoint. As in Section 8, $\{P_n^0:n\geq 1\}$ denotes the sequence of passage times (irrespective of job identity), enumerated in order of passage time start, and by the argument in Appendix 1, $P_n^0 \Rightarrow P^0$. The goal of the simulation is estimation of the quantity $r^0(f)$ of Equation (8.2.5).

Recall that to estimate $r^0(f)$, the decomposition method prescribes that we carry out the simulation of the process $\underset{\sim}{V}^0$ in random blocks defined by the successive entrances of the process to the fixed set of states U^0 defined by Equation (8.2.6). Entrances of $\underset{\sim}{V}^0$ to the set U^0 correspond to the terminations of passage times (irrespective of job identity) which occur when no other passage times are underway, and which leave a fixed configuration (z^0) of the job stack. For $k \geq 1$, γ_k^0 denotes the length in discrete time units of the kth block (returns to the set U^0) of the process $\{V_n^0 : n \geq 0\}$; also, $\delta_0^0 = 0$ and $\delta_m^0 = \gamma_1^0 + \ldots + \gamma_m^0$, $m \geq 1$.

We assume that $V^0(0) \epsilon U^0$ and for $m \geq 1$ denote by K_m^0 the number of passage times in the mth block of the process $\underset{\sim}{V}^0$. Also, we let $Y_m^0(f)$ be the sum of the quantities $f(P_n^0)$ over the passage times in the mth block of $\underset{\sim}{V}^0$. The key results of Section 8.2 leading to point estimates and confidence intervals for the quantity $r^0(f)$ are that the pairs of random variables

$$\{(Y_m^0(f), K_m^0) : m \geq 1\} \tag{9.3.3}$$

are independent and identically distributed, and that the quantity

$$r^0(f) = E_{U^0}\{Y_1^0(f)\}/E_{U^0}\{K_1^0\} , \tag{9.3.4}$$

provided that the quantity $E\{|f(P^0)|\}<\infty$. The symbol $E_{U^0}\{\cdot\}$ is an abuse of our previous notation. It connotes conditional expectation associated with starting the Markov chain $\underset{\sim}{V}^0$ in one of the states in the set U^0. The definition of the set U^0 implies that the value of this conditional expectation is independent of the particular starting state in U^0.

Given these results, from a fixed number of blocks of $\underset{\sim}{V}$, the decomposition method provides the point estimate

$$\hat{r}_n^0(f) = \bar{Y}_n^0(f)/\bar{K}_n^0 \ . \tag{9.3.5}$$

The associated confidence interval for $r^0(f)$ follows from the central limit theorem

$$\frac{n^{1/2}\{\hat{r}_n^0(f)-r^0(f)\}}{\sigma^0(f)/E_{U^0}\{K_1^0\}} \Rightarrow N(0,1) \ , \tag{9.3.6}$$

where

$$(\sigma^0(f))^2 = \text{var}\{Y_1^0(f)-r^0(f)K_1^0\} \ . \tag{9.3.7}$$

Taking f to be the identity function, we restrict attention to the quantity r^0 and consider computation of the corresponding variance constant $(\sigma^0)^2$ and related theoretical values. By the argument which leads to Equation (9.2.18), the appropriate measure of the statistical efficiency of the simulation is the quantity

$$e^0 = \left(E_{U^0}\{\alpha_1^0\}\right)^{1/2}\sigma^0/E_{U^0}\{K_1^0\} \ , \tag{9.3.8}$$

where α_1^0 is the length of a block in the continuous time process $\underset{\sim}{V}^0$.

The individual quantities required to compute this measure of efficiency are defined in terms of the successive returns of the process $\underset{\sim}{V}^0$ to a fixed set of states (U^0) rather than to a single state. Moreover, the successive entrances of $\underset{\sim}{V}^0$ to U^0 are not regeneration points for $\underset{\sim}{V}^0$. Accordingly, we cannot apply the results of Section 9.1 directly, as we did for the marked job method. Instead, we select a fixed state (designated state u^0) from the set U^0 and compute the quantity corresponding to Equation (9.3.8) for the resulting u^0-cycles. (Note that the successive entrances of the process $\underset{\sim}{V}^0$ to the fixed state u^0 are regeneration points for $\underset{\sim}{V}^0$.) The expression in Equation (9.3.8) computed for u^0-cycles is

$$e^0(u^0) = (E_{u^0}\{\alpha_1^0\})^{1/2}\sigma_0^0/E_{u^0}\{K_1^0\} ,$$

where the constant σ_0^0 (analogous to σ^0) is defined for u^0-cycles. This quantity $e^0(u^0)$ is equal to e^0. To see this, for $t\geq 0$ let $m^0(t)$ be the number of passage times (irrespective of job identity) completed in the interval $(0,t]$. In terms of simulation time, t, we have the central limit theorem

$$t^{1/2}\left[\{m^0(t)\}^{-1}\left(\sum_{i=1}^{m^0(t)} f(P_i^0)\right)-r^0(f)\right] \Big/ \left[(E_{U^0}\{\alpha_1^0\})^{1/2}\sigma^0(f)/E_{U^0}\{K_1^0\}\right] \Rightarrow N(0,1) ,$$

and, when f is the identity function, the variance constant in the denominator is the quantity e^0. There is a similar central limit theorem in terms of u^0-cycles; the numerator is the same and the variance constant in the denominator is $e^0(u^0)$. Since the numerators in these two central limit theorems are the same as are the limiting random variables (N(0,1)),

e^0 must equal $e^0(u^0)$. For a similar argument, see Propositions 5.1 and 5.6 of Crane and Iglehart (1975a).

Next we observe that the number of passage times in a u^0-cycle of the process $\underset{\sim}{V}^0$, as well as the sum of the passage times in a u^0-cycle, does not depend on the identities of the jobs in successive configurations of the job stack. It follows that rather than working with the stochastic process $\underset{\sim}{V}^0$, we can work with process $\underset{\sim}{W}=\{W(t):t\geq 0\}$ defined by

$$W(t) = (K(t),Z(t)) \ . \tag{9.3.9}$$

Here $Z(t)\epsilon D$ corresponds to the linear job stack at time t, and K(t) is the last state visited by the Markov chain $\underset{\sim}{Z}=\{Z(t):t\geq 0\}$ before jumping to Z(t). The process $\underset{\sim}{W}$ is an irreducible, positive recurrent continuous time Markov chain with a state space that is a subset of D×D. Note that in general the state space of $\underset{\sim}{W}$ is much smaller than that of $\underset{\sim}{V}^0$, and that working with the process $\underset{\sim}{W}$ is computationally advantageous.

The computations rest on the definition of two particular functions (f and g) defined on the state space of $\underset{\sim}{W}$ and taking values in the set {0,1}. We define the functions f and g in terms of functions f^0 and g^0 defined on F^0, the state space of the process $\underset{\sim}{V}^0$. We take the function f^0 to be the indicator function, 1_{S^0}, of the set S^0 which defines the starts of passage times irrespective of job identity, i.e., for $(z,n_1,\ldots,n_N,z',n_1',\ldots,n_N')\epsilon F^0$,

$$f^0(z,n_1,\ldots,n_N,z',n_1',\ldots,n_N') = 1_{S^0}(z,n_1,\ldots,n_N,z',n_1',\ldots,n_N') \ . \tag{9.3.10}$$

Thus if a passage time for some job starts when $\underset{\sim}{V}^0$ hits $(z,n_1,\ldots,n_N,z',n_1',\ldots,n_N')$, then $f^0=1$. Note that for each (z,z') in the state space of $\underset{\sim}{W}$, there exist $n_1,\ldots,n_N,n_1',\ldots$, and n_N' such that $(z,n_1,\ldots,n_N,z',n_1',\ldots,n_N')\in F^0$. For a state (z,z') of $\underset{\sim}{W}$, we define

$$f(z,z') = f^0(z,n_1,\ldots,n_N,z',n_1',\ldots,n_N') \; . \tag{9.3.11}$$

The function f is well defined since, for fixed z and z', the function f^0 is independent of its other arguments.

For an element $(z,n_1,\ldots,n_N,z',n_1',\ldots,n_N')\in F^0$, the value of the function g^0 is the number of passage times that start or are underway when the configuration of the job stack is z'. Formally, for $(z,n_1,\ldots,n_N,z',n_1',\ldots,n_N')\in F^0$, we define

$$g^0(z,n_1,\ldots,n_N,z',n_1',\ldots,n_N') = \sum_{k=1}^{N} 1_G(h(z,k)) \; . \tag{9.3.12}$$

Then, for (z,z') in the state space of $\underset{\sim}{W}$, we define

$$g(z,z') = g^0(z,n_1,\ldots,n_N,z',n_1',\ldots,n_N') \; . \tag{9.3.13}$$

The justification for using the process $\underset{\sim}{W}$ is that the number of passage times (which start and terminate) in the first u^0-cycle of $\underset{\sim}{V}^0$ is

$$\sum_{j=0}^{\kappa_1-1} f(W_j) \; , \tag{9.3.14}$$

and the sum of the passage times in the first u^0-cycle of $\underset{\sim}{V}^0$ is

$$\int_0^{\zeta_1} g(W(s))ds \; . \tag{9.3.15}$$

Here ζ_1 (respectively κ_1) is the time of the first return of the process $\underset{\sim}{W}$ (respectively the jump chain $\{W_k : k \geq 0\}$) to the fixed state w^0. The return state w^0 corresponds to the fixed state u^0 selected from the set U^0, i.e., if

$$u^0 = (z, n_1, \ldots, n_N, z^0, n_1', \ldots, n_N') ,$$

then $w^0 = (z, z^0)$.

Proposition (9.3.16) follows directly from Theorem (9.1.1).

(9.3.16) PROPOSITION. Let f be the function defined by Equation (9.3.11), and $\underline{R}$ be the transition matrix of the discrete time Markov chain $\{W_k : k \geq 0\}$. Then

$$E\{Y_1(f)\} = E\left\{\sum_{k=0}^{\kappa_1 - 1} f(W_k)\right\} = (\underline{I} - {}_0\underline{R})^{-1}\underline{f}$$

and

$$E\{(Y_1(f))^2\} = (\underline{I} - {}_0\underline{R})^{-1}\underline{h} ,$$

where κ_1 is the time of the first return to the state w^0 in $\{W_k : k \geq 0\}$ and $\underline{h} = 2\underline{f} \circ E\{Y_1(f)\} - \underline{f} \circ \underline{f}$.

From Proposition (9.3.16) we obtain the quantities $E_{u^0}\{K_1^0\}$ and $E_{u^0}\{(K_1^0)^2\}$ according to

$$E_{u^0}\{K_1^0\} = E_{w^0}\{Y_1(f)\} \tag{9.3.17}$$

and

$$E_{u^0}\{(K_1^0)^2\} = E_{w^0}\{(Y_1(f))^2\} \ . \tag{9.3.18}$$

(9.3.19) PROPOSITION. Let g be the function defined by Equation (9.3.12), and $\underline{R}$ be the jump matrix and $\underline{q}$ the vector of rate parameters for holding times in the continuous time Markov chain $\underset{\sim}{W}$. Then

$$E\{Y_1(g)\} = E\left\{\int_0^{\zeta_1} g(W(s))ds\right\} = (\underline{I}-{}_0\underline{R})^{-1}(\underline{g}\circ\underline{q}^{-1}) \ ,$$

and

$$E\{(Y_1(g))^2\} = (\underline{I}-{}_0\underline{R})^{-1}(\underline{h}\circ\underline{q}^{-1}) \ ,$$

where ζ_1 is the time of the first return to the state w^0 in $\underset{\sim}{W}$, and $\underline{h}=2\underline{g}\circ E\{Y_1(g)\}$.

Proposition (9.3.19) follows directly from Theorem (9.1.4). We use this result to obtain the quantities

$$E_{u^0}\left\{\sum_{j=1}^{K_1^0} P_j^0\right\} = E_{w^0}\{Y_1(g)\} \tag{9.3.20}$$

and

$$E_{u^0}\left\{\left(\sum_{j=1}^{K_1^0} P_j^0\right)^2\right\} = E_{w^0}\{(Y_1(g))^2\} \ . \tag{9.3.21}$$

Using the ratio formula, Equations (9.3.17) and (9.3.20) yield r^0. Analogous to Proposition (9.2.16) we have

(9.3.22) PROPOSITION. For $E\{Y_1(f)\}$ and $E\{Y_1(g)\}$ given by Propositions (9.3.16) and (9.3.19),

$$E\left\{\int_0^{\zeta_1} g(W(s))ds \sum_{k=0}^{\kappa_1-1} f(W_k)\right\} = (\underline{I}-{}_0\underline{R})^{-1}\underline{h} ,$$

where $\underline{h}=(\underline{g}\circ\underline{q}^{-1})\circ E\{Y_1(f)\}+\underline{f}\circ E\{Y_1(g)\}-(\underline{g}\circ\underline{q}^{-1})\circ\underline{f}$.

We use this result to obtain

$$E_{u^0}\left\{\left(\sum_{j=1}^{K_1^0} P_j^0\right)K_1^0\right\} = E_{w^0}\left\{\int_0^{\zeta_1} g(W(s))ds \sum_{k=0}^{\kappa_1-1} f(W_k)\right\} . \tag{9.3.23}$$

Then to compute the variance constant $(\sigma_0^0)^2$, we use the results of Equations (9.3.18), (9.3.21), and (9.3.23).

9.4. Numerical Results

We once again consider the closed network of queues of Section 5.1, and the limiting passage times P and R therein. Recall that the limiting passage time P starts when a job joins the center 1 queue upon completion of a center 2 service and terminates when the job next joins the center 2 queue. Similarly, the response time R is associated with the time between successive entrances of a job into the center 1 queue upon completion of a center 2 service.

For the passage time P, the sets A_1 and A_2 defining the starts of passage times are

$$A_1 = \{(i,N):0\le i<N\} ,$$

and

$$A_2 = \{(i,1):0<i\le N\} .$$

Similarly, the sets B_1 and B_2 defining the terminations of the passage time P are

$$B_1 = \{(i,i):0<i\le N\}$$

and

$$B_2 = \{(i-1,i):0<i\le N\} .$$

For the response time R, the sets A_1 and A_2 are the same as for the passage time P, but $B_1=A_1$ and $B_2=A_2$.

In connection with the marked job method, the process $\underset{\sim}{Y}=\{(L(t),X(t)):t\ge 0\}$, where L(t) is the last state visited by the Markov chain $\underset{\sim}{X}$ before jumping to X(t), has state space

$$F = \{(i,j,i+1,j+1):0\le j<N,\ 1\le i<N\} \cup \{(i,N,i+1,1):0\le i<N\} \cup$$

$$\{(i,j,i-1,j):0<i\le N,\ 1\le j\le N\} \cup \{(i,i,i,1):1<i\le N\} .$$

The subsets of F defining the starts and terminations of passage times for the marked job are

$$S = \{(i,N,i+1,1):0\le i<N\}$$

and

$$T = \{(i,i,i-1,i):0<i\leq N\} \ .$$

Tables 9.1 and 9.2 give theoretical values for simulation of the closed network of queues by the marked job method. Numerical results are displayed for the mean of the response time R and corresponding results for the passage time P are in parentheses. For the case of N=2 jobs (Table 9.1), the set $S=\{0,2,1,1), (1,2,1,1)\}$. With $\lambda_1=1$, $\lambda_2=0.5$, and $p=0.75$, the numerical results show that on the average 0-cycles defined by returns to the state (0,2,1,1) are twice as long as those defined by the returns to the state (1,2,1,1). Note that as expected, the quantities $\sigma^2/E_0\{M_1\}$ (as well as $e=(E_0\{\alpha_1\})^{1/2}\ \sigma/E_0\{M_1\}$) are the same for the two return states. Table 9.2 gives results for N=4 jobs. Here there are four possible return states, and for the parameter values selected, returns to the state (3,4,4,1) occur most frequently, and on the average eight times more often than returns to the state (0,4,1,1).

We now turn to the decomposition method. As we saw in Section 8.2, the process

$$\underset{\sim}{X}^0 = \{(Z(t),N^1(t),\ldots,N^N(t)):t\geq 0\}$$

has state space E^0, where

$$E^0 = \{(i,n_1,\ldots,n_N):0\leq i\leq N;\ 1\leq n_1,\ldots,n_N\leq N;\ n_i\neq n_j \text{ for } i\neq j\} \ .$$

The underlying continuous time process $\underset{\sim}{V}^0$ defined by

$$V^0(t) = (L^0(t), X^0(t)) ,$$

where $L^0(t)$ is the last state visited by the Markov chain $\underset{\sim}{X}^0$ before jumping to $X^0(t)$, has state space F^0. The subsets of F^0 defining starts and terminations of passage times are

$$S^0 = \{(i,n_1,\ldots,n_N,i+1,n_1',\ldots,n_N'):0\le i<N;\ \text{for exactly one } j^*,$$
$$n_{j^*}=N \text{ and } n'_{j^*}=1;\ 1\le n_j<N \text{ and } n_j'=n_j+1,\ j\ne j^*;\ n_k\ne n_j \text{ for } k\ne j\}$$

and

$$T^0 = \{(i,n_1,\ldots,n_N,i-1,n_1',\ldots,n_N'):0<i\le N;$$
$$1\le n_j=n_j'\le N, \text{ for } 1\le j\le N;\ n_k\ne n_j \text{ for } k\ne j\} .$$

The process $\underset{\sim}{Z}=\{Z(t):t\ge 0\}$ has state space $D=\{0,1,\ldots,N\}$, the set $D^0=\{0\}$, and the set U^0 defining blocks of the process $\underset{\sim}{V}^0$ is

$$U^0 = \{(1,n_1,\ldots,n_N,0,n_1,\ldots,n_N):1\le n_1,\ldots,n_N\le N;\ n_i\ne n_j \text{ for } i\ne j\} .$$

The state space of the stochastic process $\underset{\sim}{W}$ is

$$\{(i,i+1):0\le i\le N-1\} \cup \{(i,i-1):1\le i\le N\} ,$$

and the state $w^0=(1,0)$.

Table 9.3 gives theoretical values for simulation of the closed network of queues by the decomposition method for the mean of the passage time P. The table gives results for N=1 to N=4 jobs, and the parameter values are the same as in Tables 9.1 and 9.2. For N=2 jobs, the value of the quantity $e^0=e^0(u^0)$ of Equation (9.3.8) which measures the statistical efficiency of the decomposition method is 16.546. The corresponding value from Table 9.1 for the marked job method is 20.890. Thus, for these parameter values the decomposition method is approximately 21 percent more efficient than the marked job method. For N=4 jobs, the decomposition method is 47 percent more efficient.

Numerical results bearing on the statistical efficiency of the decomposition method for simulation of the closed network of queues appear in Table 9.4. For N=1 to N=6 jobs, the table gives theoretical values of the quantities r^0 and e^0 for three sets of parameters values. We hold the value of $\lambda_1=1$ and $p=0.75$ fixed, but vary λ_2. Table 9.5 gives a comparison of the relative efficiency (e/e^0) of the marked job and decomposition methods for the same sets of parameter values.

TABLE 9.1

Theoretical Values for the Marked Job Method.
Passage Time R (P) in Closed Network of Queues.
N=2, λ_1=1.0, λ_2=0.5, p=0.75.

Parameter	Return State of $\underset{\sim}{V}=\{V(t):t\geq 0\}$	
	(0,2,1,1)	(1,2,2,1)
$E_0\{\alpha_1\}$	24.0	12.0
$E_0\left\{\sum_{j=1}^{M_1} P_j\right\}$	28.0 (20.0)	14.0 (10.0)
$E_0\{M_1\}$	3.0 (3.0)	1.5 (1.5)
$E_0\left\{\sum_{j=1}^{M_1} P_j\right\}/E_0\{M_1\}$	9.333 (6.667)	9.333 (6.667)
σ^2	140.267 (129.067)	70.133 (64.533)
$\sigma^2/E_0\{M_1\}$	46.756 (43.022)	46.756 (43.022)
$\left(E_0\{\alpha_1\}\right)^{1/2}\ \sigma/E_0\{M_1\}$	20.890 (20.038)	20.890 (20.038)

TABLE 9.2

Theoretical Values for the Marked Job Method.
Passage Time R (P) in Closed Network of Queues.
$N=4$, $\lambda_1=1.0$, $\lambda_2=0.5$, $p=0.75$.

Parameter	Return State of $\underset{\sim}{V}=\{V(t):t\geq 0\}$			
	(0,4,1,1)	(1,4,2,1)	(2,4,3,1)	(3,4,4,1)
$E_0\{\alpha_1\}$	216.0	108.0	54.0	27.0
$E_0\left\{\sum_{j=1}^{M_1} P_j\right\}$	248.0 (196.0)	124.0 (98.0)	62.0 (49.0)	31.0 (24.5)
$E_0\{M_1\}$	15.0 (15.0)	7.5 (7.5)	3.75 (3.75)	1.875 (1.875)
$E_0\left\{\sum_{j=1}^{M_1} P_j\right\}/E_0\{M_1\}$	16.533 (13.067)	16.533 (13.067)	16.533 (13.067)	16.533 (13.067)
σ^2	2319.479 (2343.808)	1159.739 (1171.904)	579.870 (585.952)	289.934 (292.976)
$\sigma^2/E_0\{M_1\}$	154.632 (156.254)	154.632 (156.254)	154.632 (156.254)	154.632 (156.254)
$\left(E_0\{\alpha_1\}\right)^{1/2}\ \sigma/E_0\{M_1\}$	50.563 (50.827)	50.563 (50.827)	50.563 (50.827)	50.563 (50.827)

TABLE 9.3

Theoretical Values for the Decomposition Method.
Passage Time P in Closed Network of Queues.
$\lambda_1=1.0$, $\lambda_2=0.5$, $p=0.75$.

Parameter	N=1	N=2	N=3	N=4
$E_{u^0}\{\alpha_1^0\}$	6.0	24.0	30.0	62.0
$E_{u^0}\left\{\sum_{j=1}^{K_1^0} P_j^0\right\}$	4.0	20.0	68.0	196.0
$E_{u^0}\{K_1^0\}$	1.0	3.0	7.0	15.0
$E_{u^0}\left\{\sum_{j=1}^{K_1^0} P_j^0\right\}/E_{u^0}\{K_1^0\}$	4.0	6.667	9.714	13.067
$(\sigma_0^0)^2$	16.0	176.0	1023.673	4317.227
$(\sigma_0^0)^2/E_{u^0}\{K_1^0\}$	16.0	58.667	146.249	287.815
$\left(E_{u^0}\{\alpha_1^0\}\right)^{1/2} \sigma_0^0/E_{u^0}\{K_1^0\}$	9.798	16.546	25.035	34.491

TABLE 9.4

Statistical Efficiency of the Decomposition Method.
Passage Time P in Closed Network of Queues.

	$p = 0.75$, $\lambda_1 = 1.0$, $\lambda_2 = 0.125$		$p = 0.75$, $\lambda_1 = 1.0$, $\lambda_2 = 0.25$		$p = 0.75$, $\lambda_1 = 1.0$, $\lambda_2 = 0.5$	
N	r^0	e^0	r^0	e^0	r^0	e^0
1	4.0	13.856	4.0	11.314	4.0	9.798
2	5.333	19.956	6.0	17.664	6.667	16.546
3	6.286	27.380	8.0	26.128	9.714	25.035
4	6.933	35.189	10.0	36.606	13.067	34.491
5	7.355	42.597	12.0	49.107	16.645	44.296
6	7.619	49.068	14.0	63.645	20.381	54.021

TABLE 9.5

Relative Efficiency of the Marked Job and Decomposition Methods.
Passage Time P in Closed Network of Queues.

N	$p = 0.75$ $\lambda_1 = 1.0$ $\lambda_2 = 0.125$	$p = 0.75$ $\lambda_1 = 1.0$ $\lambda_2 = 0.25$	$p = 0.75$ $\lambda_1 = 1.0$ $\lambda_2 = 0.5$
1	1.0	1.0	1.0
2	1.190	1.189	1.211
3	1.259	1.265	1.351
4	1.285	1.294	1.474
5	1.301	1.301	1.597
6	1.317	1.297	1.725

10.0. NETWORKS WITH MULTIPLE JOB TYPES

We have considered in previous sections the problem of simulating closed and finite capacity open networks of queues, respectively, for general characteristics of passage times. Under consideration here are networks with multiple job types and the estimation of individual and joint characteristics of passage times over the several job types. The type of a job may influence its routing through the network as well as its service requirements at each center. For expository convenience, we assume that there are only two job types in the network and we mark one job of each type. By tracking these two jobs, we are able to produce from a single replication confidence intervals for a variety of passage time characteristics. The estimation method of this section can also be applied to networks with only a single job type; the result is an alternative scheme to that proposed in Section 7.

10.1. Preliminaries

We consider closed networks of queues with a finite number of jobs, N, of two types, and assume that there are N_1 [respectively N_2] jobs of type 1 [respectively type 2] with $N_1+N_2=N$. In each network there are a finite number of service centers, s, and a finite number of job classes, c. All jobs retain their job type, but may change class as they traverse the network. (Think of type 1 jobs as cubes and type 2 jobs as spheres, and let job classes correspond to different colors. Then we permit jobs to change color, but not shape.) Upon completion of service at center i, a type ν job of class j goes to center k and changes to class ℓ with probability $p^{(\nu)}_{ij,k\ell}$. We assume that for $\nu=1,2$, $\underline{P}^{(\nu)}=\{p^{(\nu)}_{ij,k\ell}:1\le i,k\le s,\ 1\le j,\ell\le c\}$ is a given irreducible Markov matrix.

The service times and service discipline at each service center are as in Section 7 with the exception that they may also depend on job type. We briefly review the situation. At each service center jobs queue and receive service according to a fixed priority scheme among classes and types, which scheme can vary from center to center. Each center operates as a single server, processing jobs of a fixed type and class according to a fixed service discipline. All service times in the network are mutually independent, and at each center have a distribution with a Cox-phase representation with parameters which may depend on the service center, type and class of job being serviced, and the "state" of the entire system. (As usual, we exclude zero service times occurring with positive probability.) A job in service may or may not be preempted (according to a fixed procedure for each center) if another job of higher priority joins the queue at the center.

We restrict the present discussion to networks in which all service times are exponentially distributed, and deal with distributions having a Cox-phase representation in the usual way by the method of stages. To characterize the state of the system at time t, we let $S_i(t)$ denote the (type, class) pair of the job receiving service at center i at time t, $i=1,2,\ldots,s$. If there are no jobs at center i at time t, we set $S_i(t)=(0,0)$. We denote by $j_1(i),\ldots,j_{k(i)}(i)$ the (type, class) pairs served at center i ordered by decreasing priority, and let

$C^{(i)}_{j_1}(t),\ldots,C^{(i)}_{j_{k(i)}}(t)$ denote the number of jobs in queue at time t of the various (type, class) pairs served at center i. We mark one job of each of the two types in order to measure their passage or response times. As in previous sections, we view the N jobs as being completely ordered in a linear stack, and let the vector Z(t) be given by:

$$Z(t) = (C^{(1)}_{j_{k(1)}}(t),\ldots,C^{(1)}_{j_1}(t),S_1(t);\ldots;\\ C^{(s)}_{j_{k(s)}}(t),\ldots,C^{(s)}_{j_1}(t),S_s(t)) . \qquad (10.1.1)$$

The linear job stack again corresponds to the order of components in the vector Z(t) after ignoring any zero components. Within a (type, class) pair at a center, jobs waiting appear in the job stack in the order of their arrival in the center, the latest to arrive being closest to the top of the stack. Let $N_\nu(t)$ ($\nu=1,2$) denote the position from the top in this job stack of the type ν marked job. Then for $t\geq 0$, the state vector of the network is

$$X(t) = (Z(t),N_1(t),N_2(t)) . \qquad (10.1.2)$$

Under the exponential service time and Markovian routing assumptions, the process $\underset{\sim}{X}=\{X(t):t\geq 0\}$ is an irreducible continuous time Markov chain with finite state space E.

10.2. Simulation for Passage Times

We specify the passage (or response) times for the two types of jobs by eight subsets of E: $A_1^{(\nu)}$, $A_2^{(\nu)}$, $B_1^{(\nu)}$, $B_2^{(\nu)}$, for $\nu=1,2$. The sets $A_1^{(\nu)}$ and $A_2^{(\nu)}$ [respectively $B_1^{(\nu)}$, $B_2^{(\nu)}$] determine when to start [respectively stop] the clock measuring a particular passage time for the type ν marked job. Denoting the jump times of the process $\underset{\sim}{X}$ by $\{\tau_n:n\geq 0\}$, for k,n≥1 we require that the sets $A_1^{(\nu)}$, $A_2^{(\nu)}$, $B_1^{(\nu)}$ and $B_2^{(\nu)}$ satisfy

$$\text{if } X(\tau_{n-1})\in A_1^{(\nu)},\ X(\tau_n)\in A_2^{(\nu)},\ X(\tau_{n-1+k})\in A_1^{(\nu)} \text{ and } X(\tau_{n+k})\in A_2^{(\nu)} ,$$

$$\text{then } X(\tau_{n-1+m})\in B_1^{(\nu)} \text{ and } X(\tau_{n+m})\in B_2^{(\nu)} \text{ for some } 0<m\leq k ;$$

and

$$\text{if } X(\tau_{n-1})\in B_1^{(\nu)},\ X(\tau_n)\in B_2^{(\nu)},\ X(\tau_{n-1+k})\in B_1^{(\nu)} \text{ and } X(\tau_{n+k})\in B_2^{(\nu)} ,$$

$$\text{then } X(\tau_{n-1+m})\in A_1^{(\nu)} \text{ and } X(\tau_{n+m})\in A_2^{(\nu)} \text{ for some } 0\leq m<k .$$

Also, in terms of the jump times of $\underset{\sim}{X}$, we define four sequences of random times: $\{S_j^{(\nu)}:j\geq 0\}$ and $\{T_j^{(\nu)}:j\geq 1\}$, for $\nu=1,2$. The start [respectively termination] time of the jth passage time for the type ν marked job is denoted by $S_{j-1}^{(\nu)}$ [respectively $T_j^{(\nu)}$]. Formally, for $\nu=1,2$ we have

$$S_j^{(\nu)} = \inf\{\tau_n\geq T_j^{(\nu)}:X(\tau_n)\in A_2^{(\nu)},\ X(\tau_{n-1})\in A_1^{(\nu)}\},\ j\geq 1$$

and

$$T_j^{(\nu)} = \inf\{\tau_n>S_{j-1}^{(\nu)}:X(\tau_n)\in B_2^{(\nu)},\ X(\tau_{n-1})\in B_1^{(\nu)}\},\ j\geq 1 ,$$

where $S_0^{(\nu)}$ is the start of the first passage time for the type ν marked job after time zero. The jth passage time for the type ν marked job is $P_j^{(\nu)}=T_j^{(\nu)}-S_{j-1}^{(\nu)}$, j≥1. Note that the definition of these times is as in Section 7. For response times of type ν jobs, $A_1^{(\nu)}=B_1^{(\nu)}$, $A_2^{(\nu)}=B_2^{(\nu)}$, and $S_j^{(\nu)}=T_j^{(\nu)}$ for all j≥1.

Let L(t) denote the last state visited by the Markov chain $\underset{\sim}{X}$ before jumping to X(t), and for t≥0 set

$$V(t) = (L(t),X(t)) . \tag{10.2.1}$$

The process $\underset{\sim}{V}=\{V(t):t\geq 0\}$ has a state space F consisting of all pairs of states (i,j), i,jϵE, for which a transition in $\underset{\sim}{X}$ from state i to state j can occur with positive probability. In general, of course, the size of the state space F is larger than that of E. The "Q-matrix" used in generating the Markov chain $\underset{\sim}{V}$ can be obtained easily from that for $\underset{\sim}{X}$. Since $\underset{\sim}{X}$ is an irreducible, positive recurrent Markov chain, so is $\underset{\sim}{V}$. Clearly, the entrance times of $\underset{\sim}{V}$ to a state (i,j)ϵF correspond to the times of transition in $\underset{\sim}{X}$ from state i to state j. For a type ν job, we define two subsets of F according to:

$$S^{(\nu)} = \{(i,j)\epsilon F: i\epsilon A_1^{(\nu)},\ j\epsilon A_2^{(\nu)}\}$$

$$T^{(\nu)} = \{(i,j)\epsilon F: i\epsilon B_1^{(\nu)},\ j\epsilon B_2^{(\nu)}\} .$$

Thus the entrances of $\underset{\sim}{V}$ to $S^{(\nu)}$ [respectively $T^{(\nu)}$] correspond to the start [respectively termination] times of passage times for the type ν marked job. Of course for response times of a type ν job, $S^{(\nu)}=T^{(\nu)}$.

The argument employed in Appendix 1 shows that for $\nu=1,2$, the sequence $P_n^{(\nu)}$ converges in distribution to a random variable $P^{(\nu)}$. Moreover, the sequence of passage times of type ν jobs (irrespective of job identity) in the order of start (or termination) also converges in distribution to $P^{(\nu)}$. Our concern is with the estimation of characteristics associated with these limiting passage times.

<u>Estimation of $E\{R^{(1)}\}$ and $P\{R^{(1)}\le x\}$</u>

Using the process $\underset{\sim}{V}$ defined by Equations (10.1.1), (10.1.2) and (10.2.1), we consider first the estimation of characteristics of the limiting response time, $R^{(1)}$, of a type 1 job. For this estimation problem, of course, it is not necessary to mark a type 2 job. Since $R^{(1)}$ is a response time, $S^{(1)}=T^{(1)}$. We select a fixed state of $S^{(1)}$, which for convenience we designate state 0, and assume that $\underset{\sim}{V}(0)=0$.

Suppose that we wish to estimate $E\{R^{(1)}\}$. The successive entrances of $\underset{\sim}{V}$ to $S^{(1)}$ constitute the starts and terminations of response times of the type 1 marked job. Let $R_n^{(1)}$ $(n\ge 0)$ denote the time between the nth and (n+1)st entrances to $S^{(1)}$, with the 0th entrance to $S^{(1)}$ occurring at $t=0$. Also, let $\{V_n:n\ge 0\}$ denote the embedded jump chain associated with $\underset{\sim}{V}$. The random times $\{\alpha_n:n\ge 1\}$ and $\{\gamma_n:n\ge 1\}$ denote the lengths of the successive 0-cycles (successive returns to the fixed state 0) for $\underset{\sim}{V}$ and $\{V_n:n\ge 0\}$, respectively. Then the number of response times for the type 1 marked job in the first 0-cycle of $\underset{\sim}{V}$ is

$$N_1^{(1)} = \sum_{j=0}^{\delta_1-1} 1_{\{V_n\in S^{(1)}\}}$$

where $\delta_0=0$ and $\delta_m=\gamma_1+\ldots+\gamma_m$, $m\ge 1$. The sum of the response times in that cycle is simply

$$\alpha_1 = \sum_{n=1}^{N_1^{(1)}} R_n^{(1)} .$$

We denote the analogous quantities in the kth 0-cycle by $N_k^{(1)}$ and α_k. The fact that $\underset{\sim}{V}$ is a regenerative process, together with a renewal argument (cf. Appendix 2) establishes

(10.2.2) PROPOSITION. The pairs of random variables $\{(\alpha_k,N_k^{(1)}):k\geq 1\}$ are independent and identically distributed. Provided that $E\{R^{(1)}\}<\infty$,

$$E\{R^{(1)}\} = E\{\alpha_1\}/E\{N_1^{(1)}\} .$$

At this point, the arguments of the standard regenerative method hold and, based on n cycles, we can construct the point estimate $\overline{\alpha}_n/\overline{N}_n^{(1)}$ and (provided that an estimate is available for σ^2, the variance of $\alpha_1-E\{R^{(1)}\}N_1^{(1)}$) an associated confidence interval for $E\{R^{(1)}\}$. The confidence interval is obtained from the central limit theorem

$$\frac{n^{1/2}[\overline{\alpha}_n/\overline{N}_n^{(1)}-E\{R^{(1)}\}]}{\sigma/E\{N_1^{(1)}\}} \Rightarrow N(0,1) ,$$

Here $\overline{\alpha}_n=(\alpha_1+\ldots+\alpha_n)/n$ and $\overline{N}_n^{(1)}=(N_1^{(1)}+\ldots+N_n^{(1)})/n$.

If we are interested in the distribution function, $P\{R^{(1)}\leq x\}$ of $R^{(1)}$, we proceed as above, but define in addition the i.i.d. sequence of random variables $\{Y_k:k\geq 1\}$, where, for example,

$$Y_1 = \sum_{n=1}^{N_1^{(1)}} 1_{\{R_n^{(1)}\leq x\}} .$$

Then the point estimate of $P\{R^{(1)}\leq x\}$ is just $\overline{Y}_n/\overline{N}_n^{(1)}$, and we obtain confidence intervals in the usual way.

Estimation of $E\{R^{(1)}\}$ and $E\{R^{(2)}\}$

Now suppose that we wish to estimate the expected passage time for type 2 jobs, $E\{R^{(2)}\}$, as well as $E\{R^{(1)}\}$. Response times for the type 2 marked job start and terminate at the entrance times of $\underset{\sim}{V}$ to the set $S^{(2)}=T^{(2)}$. Let $N_k^{(2)}$ denote the number of entrances to $S^{(2)}$ of $\underset{\sim}{V}$ in the kth 0-cycle. For example, in the first 0-cycle

$$N_1^{(2)} = \sum_{n=0}^{\delta_1-1} 1_{\{V_n \in S^{(2)}\}} .$$

Although we are able to begin the simulation at the start of a response time for the type 1 marked job, in general a response time for the type 2 marked job is underway at time t=0. Similarly, at the end of a 0-cycle, a response time for the type 1 marked job terminates, but a response time for the type 2 marked job is still underway. After n 0-cycles, $N_1^{(2)}+\ldots+N_n^{(2)}$ response times for the type 2 marked job have started and the sum of these response times is approximately $\alpha_1+\ldots+\alpha_n$. The error in this approximation is due to the partial response time at t=0 which is not counted in $N_1^{(2)}+\ldots+N_n^{(2)}$ and the last response time which is counted, but does not terminate before the end of the nth 0-cycle. Since the point estimates and confidence intervals here are based on large sample theory (strong laws and central limit theorems), these errors are negligible for n large. In fact, the errors due to the two response times at t=0 and at the end of the simulation run compensate for each other. Consequently, we have

<u>(10.2.3) PROPOSITION</u>. The pairs of random variables $\{(\alpha_k,N_k^{(2)}):k\geq 1\}$ are independent and identically distributed. Provided that $E\{R^{(2)}\}<\infty$,

$$E\{R^{(2)}\} = E\{\alpha_1\}/E\{N_1^{(2)}\} .$$

In the presence of Proposition (10.2.3), the point estimate of $E\{R^{(2)}\}$ is $\bar{\alpha}_n/\bar{N}_n^{(2)}$, and we can use the standard regenerative method to obtain a confidence interval.

<u>Estimation of $E\{R^{(1)}\}-E\{R^{(2)}\}$</u>

Suppose now that we wish to estimate $r^{(1)}-r^{(2)}$, where $r^{(1)}=E\{R^{(1)}\}$ and $r^{(2)}=E\{R^{(2)}\}$. We can take as a point estimate the quantity $(\bar{\alpha}_n/\bar{N}_n^{(1)})-(\bar{\alpha}_n/\bar{N}_n^{(2)})$, but need a bivariate central limit theorem in order to produce a confidence interval. To this end, we let

$$Z_k^{(\nu)} = \alpha_k - r^{(\nu)} N_k^{(\nu)}$$

and

$$\underline{Z}_k = \begin{pmatrix} Z_k^{(1)} \\ Z_k^{(2)} \end{pmatrix}$$

for $k\geq 1$. (We take all our vectors to be column vectors.) The random vectors $\{\underline{Z}_k : k\geq 1\}$ are i.i.d. since each $\underline{Z}_k$ is only a function of the kth 0-cycle. Furthermore, Equations (10.1.4) and (10.1.5) imply that $E\{\underline{Z}_k\}=\underline{0}$. Denoting the transpose of $\underline{Z}_k$ by $\underline{Z}_k'$, let $\underline{\Sigma}=E\{\underline{Z}_k\underline{Z}_k'\}=\{\sigma_{ij}\}$ be the covariance matrix of the $\underline{Z}_k$'s. Assuming that the elements of $\underline{\Sigma}$ are finite, we have the central limit theorem

$$n^{-1/2} \sum_{k=1}^{n} \underline{Z}_k \Rightarrow N(\underline{0},\underline{\Sigma}) \ , \qquad (10.2.4)$$

where $N(\underline{0},\underline{\Sigma})$ is a multivariate normal random variable with zero mean vector and covariance matrix $\underline{\Sigma}$. We can rewrite Equation (10.2.4) in the form

$$n^{1/2}\begin{bmatrix} (\bar{N}_n^{(1)}/E\{N_1^{(1)}\})E\{N_1^{(1)}\}[\{(\bar{\tau}_n/\bar{N}_n^{(1)})-r^{(1)}\}] \\ (\bar{N}_n^{(2)}/E\{N_1^{(2)}\})E\{N_1^{(2)}\}[\{(\bar{\alpha}_n/\bar{N}_n^{(2)})-r^{(2)}\}] \end{bmatrix} \Rightarrow N(\underline{0},\underline{\Sigma}) \ . \qquad (10.2.5)$$

Since $E\{N_1^{(\nu)}\}/\overline{N}_n^{(\nu)} \Rightarrow 1$, we can use Lemma (2.1.8) to conclude that the factors $(\overline{N}_n^{(\nu)}/E\{N_1^{(\nu)}\})$ in Equation (10.2.5) can be dropped. With these factors removed, again apply Lemma (2.1.8) with the mapping h given by

$$h(x_1,x_2) = (x_1/E\{N_1^{(1)}\},\ x_2/E\{N_1^{(2)}\})$$

to obtain

$$n^{1/2}\begin{bmatrix}(\overline{\alpha}_n/\overline{N}_n^{(1)})-r^{(1)}\\(\overline{\alpha}_n/\overline{N}_n^{(2)})-r^{(2)}\end{bmatrix} \Rightarrow N(\underline{0},\underline{B\Sigma B'})\ , \tag{10.2.6}$$

where

$$\underline{B} = \begin{bmatrix}1/E\{N_1^{(1)}\} & 0\\0 & 1/E\{N_1^{(2)}\}\end{bmatrix}.$$

Note that from Equation (10.2.6) we could construct a simultaneous confidence interval for $(r^{(1)},r^{(2)})$. Finally, a third application of Lemma (2.1.8), this time using $h(x_1,x_2)=x_1-x_2$, yields

(10.2.7) PROPOSITION. Provided that σ_{11}, σ_{12}, $\sigma_{22}<\infty$,

$$(n^{1/2}/\sigma)[\{(\overline{\alpha}_n/\overline{N}_n^{(1)})-(\overline{\alpha}_n/N_n^{(2)})\}-(r^{(1)}-r^{(2)})] \Rightarrow N(0,1) \tag{10.2.8}$$

where

$$\sigma^2 = \sigma_{11}/E^2\{N_1^{(1)}\} + \sigma_{22}/E^2\{N_1^{(2)}\} - 2\sigma_{12}/(E\{N_1^{(1)}\}E\{N_1^{(2)}\})\ .$$

We can use the central limit theorem of Equation (10.2.8) to construct a confidence interval for $r^{(1)}-r^{(2)}$, provided that an estimate for the constant σ is available. Using the classical method, we can estimate σ from the sequence of observations taken in the n 0-cycles of the process $\underset{\sim}{V}$. This estimate for σ appears in Appendix 3.

A special case of the situation just discussed is when the two types of jobs are the same; then there is only one job type, but we elect to mark two jobs. Let $r^{(1)}=r^{(2)}=r$, $\hat{r}_n^{(\nu)}=\bar{\alpha}_n/\bar{N}_n^{(\nu)}$, and $\underline{\hat{r}}_n=(\hat{r}_n^{(1)},\hat{r}_n^{(2)})$. Then we can use the method of multiple estimates of Heidelberger (1977) applied to Equation (10.2.6). For any vector $\underline{\beta}=(\beta_1,\beta_2)$ with $\beta_1+\beta_2=1$, we have

$$(n^{1/2}/\sigma(\underline{\beta}))\ (\underline{\beta}'\underline{\hat{r}}_n-r) \Rightarrow N(0,1) ,$$

where $\sigma^2(\underline{\beta})=\underline{\beta}'(\underline{B\Sigma B'})\underline{\beta}$. Next we select that value of $\underline{\beta}$, call it $\underline{\beta}^*$, which minimizes $\sigma^2(\underline{\beta})$ subject to $\underline{\beta}'\underline{e}=1$, where $\underline{e}=(1,1)$. It turns out that $\underline{\beta}^*$ is given by

$$\underline{\beta}^* = \{1/(\underline{e}'(\underline{B\Sigma B'})^{-1}\underline{e})\}(\underline{B\Sigma B'})^{-1}\underline{e}$$

and

$$\sigma^2(\underline{\beta}^*) = 1/\{\underline{e}'(\underline{B\Sigma B'})^{-1}\underline{e}\} . \qquad (10.2.9)$$

Since $\underline{\beta}=(1,0)$ is one possible value of $\underline{\beta}$, using $\underline{\beta}^*$ is guaranteed to yield a variance reduction over that obtained by marking just one job. Again, of course, we must estimate the variance $\sigma^2(\underline{\beta}^*)$ given in Equation (10.2.9) from the observations recorded.

Estimation of $P\{R^{(1)}\le x\}-P\{R^{(2)}\le x\}$

Finally, we consider the estimation of $P\{R^{(1)}\le x\}-P\{R^{(2)}\le x\}$ for a given value of x. This is the most difficult of the problems for networks with multiple job types that we treat. Since the value of x is fixed throughout the discussion, in general we suppress in our notation the dependence of x. Again we form 0-cycles based on the response times for the type 1 marked job. Here, however, when a 0-cycle ends, we do not know whether the response time for the type 2 marked job in progress will be less than or equal to x. Thus, with respect to the response times for the type 2 marked job, the 0-cycles used previously do not create the i.i.d. cycles needed to establish a central limit theorem. Instead, we form new cycles by grouping together a random number of consecutive 0-cycles. Let $t_i = \alpha_1 + \ldots \alpha_i$, $i \geq 1$. Then let s_i be the start time of the response time for the type 2 marked job underway at the conclusion of the ith 0-cycle. We assume that $\underset{\sim}{V}(0)=0$ and regard the value of the response time for the type 2 marked job underway at the start of the simulation to be greater than (the fixed) x. We do this so that the start of the simulation corresponds to the beginning of one of the new "super-cycles" we are constructing. Defining a random variable γ according to

$$\gamma = \inf\{i \geq 1 : t_i - s_i > x\} ,$$

the length of the first super-cycle is simply $\alpha_1 + \alpha_2 + \ldots + \alpha_\gamma$, and the number of response times for the type ν marked job started in this super-cycle is $n_1^{(\nu)} = N_1^{(\nu)} + \ldots + N_\gamma^{(\nu)}$. Successive super-cycles are defined in an analogous fashion. For $k \geq 1$, we define $Y_k^{(\nu)}$ to be the number of response times terminating in the kth super-cycle which are less than or equal to x; e.g.,

$$Y_1^{(\nu)} = \sum_{k=0}^{n_1^{(\nu)}-1} 1_{\{R_k^{(\nu)} \leq x\}} .$$

Observe that by the definition of a super-cycle, the first response time of the type 2 marked job terminating within a super-cycle must be greater than x. Thus we have

(10.2.10) PROPOSITION. The random variables $\{Y_k^{(2)}:k\geq 1\}$ are independent and identically distributed.

Of course, the $Y_k^{(1)}$'s are i.i.d. also, as are the $n_k^{(1)}$'s and $n_k^{(2)}$'s. We can now form the bivariate central limit theorem analogous to Equation (10.2.6), namely

$$n^{1/2}\begin{bmatrix}(\bar{Y}_n^{(1)}/\bar{n}_n^{(1)})-P\{R^{(1)}\leq x\}\\ (\bar{Y}_n^{(2)}/\bar{n}_n^{(2)})-P\{R^{(2)}\leq x\}\end{bmatrix} \Rightarrow N(\underline{0},\underline{B}(x)\underline{\Sigma}(x)\underline{B}'(x)) ,$$

where

$$\underline{B}(x) = \begin{bmatrix}1/E\{n_1^{(1)}\} & 0\\ 0 & 1/E\{n_1^{(2)}\}\end{bmatrix}$$

and

$$\underline{\Sigma}(x) = \{\sigma_{ij}(x)\}$$

with

$$\underline{\sigma}_{ij}(x) = E\{[Y_1^{(i)}-n_1^{(i)}P\{R^{(i)}\leq x\}][Y_1^{(j)}-n_1^{(j)}P\{R^{(j)}\leq x\}]\} .$$

Finally, by the same argument used in Proposition (10.2.7), we obtain

(10.2.11) PROPOSITION. Provided that $\sigma_{11}(x)$, $\sigma_{12}(x)$, $\sigma_{22}(x)<\infty$,

$$(n^{1/2}/\sigma(x))[(\bar{Y}_n^{(1)}/\bar{n}_n^{(1)}-\bar{Y}^{(2)}/\bar{n}_n^{(2)})-(P\{R^{(1)}\leq x\}-P\{R^{(2)}\leq x\})] \Rightarrow N(0,1) , \quad (10.2.12)$$

where

$$\sigma^2(x) = \sigma_{11}(x)/E^2\, n_1^{(1)}\} + \sigma_{22}(x)/E^2\{n_1^{(2)}\} - 2\sigma_{12}(x)/(E\{n_1^{(1)}\}E\{n_1^{(2)}\}) .$$

We can estimate the quantity $\sigma(x)$ from the observations in the n super-cycles using the classical method; see Appendix 3. Then we construct confidence intervals for $P\{R^{(1)} \leq x\} - P\{R^{(2)} \leq x\}$ from Equation (10.2.12) in the usual way.

The discussion in this section has concentrated on problems associated with the estimation of characteristics of response times for the two types of jobs. The estimation of characteristics of two passage times, or one response time and one passage time, is in general easier. This is because there is the possibility of forming, from 0-cycles based on one type of job, super-cycles which terminate when no passage time of the other type of job is underway.

We have considered explicitly only the case of two job types. The estimation methods of this section apply equally well to networks having more than two job types. The state space which results from the augmentation of the vector $X(t)$ (by components to track a marked job of each of the job types) is of course larger.

10.3. Example and Numerical Results

To illustrate the technique of the previous section for estimation of response times, we consider a simple closed network of queues having two types of jobs and two service centers; see Figure 10.1. There are N jobs in the network, N_1 jobs of type 1 and N_2 jobs of type 2. After completion of service in center 1, a type ν job joins the queue at center 1 (with probability $p^{(\nu)}$) or joins the queue in center 2 (with probability $1-p^{(\nu)}$).

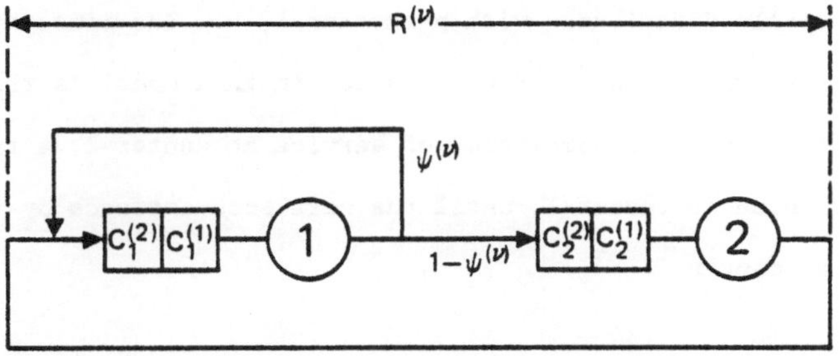

(i) Services at centers 1 and 2 are not interruptable

(ii) Routing for type ν jobs determined by binary valued variable $\psi^{(\nu)}$

(iii) Type 1 jobs have non-preemptive priority over type 2 jobs

Figure 10.1. Closed network of queues with two job types

After completion of service at center 2, jobs join the queue at center 1. At both service centers, type 1 jobs have nonpreemptive priority over type 2 jobs. Jobs of the same type at either of the centers receive service in order of their arrival at the center. We assume that all service times are mutually independent; jobs of type ν at center i receive service which is exponentially distributed with parameter $\lambda_i^{(\nu)}$. The limiting response time $R^{(\nu)}$ for type ν jobs that we consider in this model is the time measured from when upon completion of service at center 2, a type ν job enters the queue at center 1, until the next such entrance by the job into the queue at center 1.

In this model, there are two job classes, class 1 jobs at center 1 and class 2 jobs at center 2. Each center sees both job types, but only one job class. The irreducible Markov routing matrices $\underline{P}^{(\nu)}$ are of the form

$$\underline{P}^{(\nu)} = \begin{bmatrix} p^{(\nu)} & 1-p^{(\nu)} \\ 1 & 0 \end{bmatrix}.$$

Since type 1 jobs have priority over type 2 jobs at both centers, the (type, class) pairs ordered by decreasing priority are $j_1(i)=(1,i)$ and $j_2(i)=(2,i)$, $i=1,2$. For this model, it is sufficient to take as the component $S_i(t)$ in the vector $Z(t)$ the type of job in service at center i at time t, rather than the (type, class) pair. Then we can define the vector $Z(t)$ as

$$Z(t) = (C_1^{(2)}(t), C_1^{(1)}(t), S_1(t), C_2^{(2)}(t), C_2^{(1)}(t), S_2(t)) ,$$

where, for $i=1,2$ and $\nu=1,2$,

$C_i^{(\nu)}(t)$ = number of type ν jobs in queue at center i at time t ,

and

$S_i(t)$ = type of job in service at center i at time t

0 if center i is idle at time t .

Letting $N_\nu(t)$ (ν=1,2) denote the position from the top of the type ν marked job in the linear job stack, for t≥0 the state vector for this model is

$$X(t) = (Z(t),N_1(t),N_2(t)) \ .$$

Letting L(t) denote the last state visited by the Markov chain $\underset{\sim}{X}=\{X(t):t\geq 0\}$ before jumping to X(t), the vector V(t) is

$$V(t) = (L(t),X(t)) \ .$$

For N=2 jobs, the state space E of the process $\{X(t):t\geq 0\}$ has six states and is

$$E = \{(0,0,0,1,0,1,2,1),\ (0,0,0,0,1,2,1,2),\ (0,0,1,0,0,2,1,2)\}$$
$$\cup \{(0,0,2,0,0,1,2,1),\ (1,0,1,0,0,0,2,1),\ (0,1,2,0,0,0,1,2,)\} \ .$$

The subsets $A_1^{(1)}$ and $A_2^{(1)}$ of E defining the start of response times for the type 1 marked job are

$$A_1^{(1)} = \{(0,0,0,1,0,1,2,1),\ (0,0,2,0,0,1,2,1)\}$$

and

$$A_2^{(1)} = \{(0,0,1,0,0,2,1,2),\ (0,1,2,0,0,0,1,2)\} \ .$$

Similarly, the subsets $A_1^{(2)}$ and $A_2^{(2)}$ of E defining the start of response times for the type 2 marked job are

$$A_1^{(2)} = \{(0,0,0,0,1,2,1,2),\ (0,0,1,0,0,2,1,2)\}$$

and

$$A_2^{(2)} = \{(0,0,2,0,0,1,2,1),\ (1,0,1,0,0,0,2,1)\} .$$

Since $R^{(1)}$ and $R^{(2)}$ are response times, $B_1^{(\nu)}=A_1^{(\nu)}$ and $B_2^{(\nu)}=A_2^{(\nu)}$, $\nu=1,2$; see Figure 10.2. It is easy to check that the state space F of the process $\{V(t):t\geq 0\}$ has nine states. The subsets $S^{(\nu)}$ of F defining the starts of response times for the type ν marked job are

$$S^{(1)} = \{(0,0,0,1,0,1,2,1,0,0,1,0,0,2,1,2)\}$$
$$\cup\ \{(0,0,2,0,0,1,2,1,0,1,2,0,0,0,1,2)\}$$

and

$$S^{(2)} = \{(0,0,0,0,1,2,1,2,0,0,2,0,0,1,2,1)\}$$
$$\cup\ \{(0,0,1,0,0,2,1,2,1,0,1,0,0,0,2,1)\} ,$$

respectively; see Figure 10.3. Here we use the enumeration of the six states of E given in Figure 10.2. Thus, e.g., (1,3) denotes the state $(0,0,0,1,0,1,2,1,0,0,1,0,0,2,1,2)\in F$.

Simulation results for this model for $p^{(1)}=p^{(2)}=p=0.75$, $\lambda_1^{(1)}=\lambda_1^{(2)}=\lambda_1=1$ and $\lambda_2^{(1)}=\lambda_2^{(2)}=\lambda_2=0.5$, with $N=2$, appear in Tables 10.1-10.4. With these parameter values, there is one type 1 job and one type 2 job. The routing and service requirements of the two job types are the same; the two jobs differ only with respect to the nonpreemptive priority given (at each center) to the type 1 job. The simulation used the congruential uniform

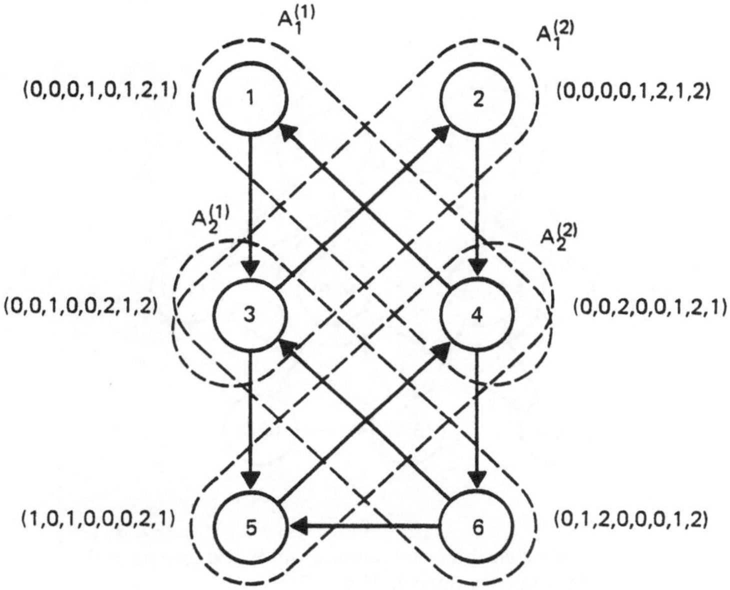

Figure 10.2. State transitions in Markov chain **X** and subsets of E for response times $R^{(1)}$ and $R^{(2)}$

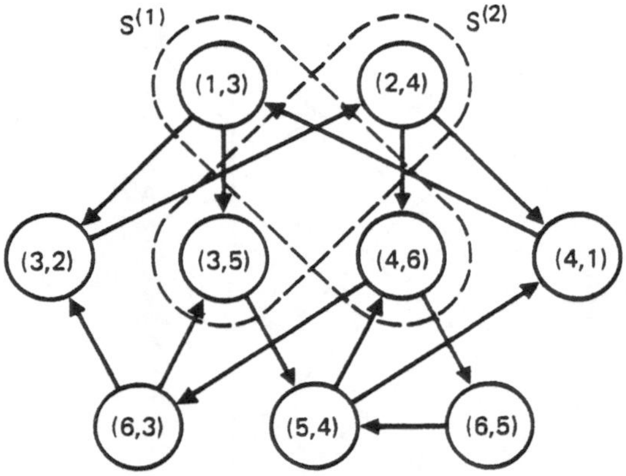

Figure 10.3. State transitions in Markov chain V and subsets of F for response times $R^{(1)}$ and $R^{(2)}$

random number generator described by Lewis, Goodman, and Miller (1969), with exponential service times obtained by logarithmic transformation of the uniform random numbers. Independent streams of exponential random numbers (obtained from different seeds) were used to generate individual exponential holding time sequences.

For the simulation results of Tables 10.1-10.4, the return state defining 0-cycles of the response time for the type 1 job is the state (0,0,2,0,0,1,2,1,0,1,2,0,0,0,1,2). This corresponds to a response time for the type 1 (marked) job starting when the type 2 (marked) job is in service at center 1. Table 10.1 summarizes results of the simulation and reports point estimates and 90 percent confidence intervals for the quantities $E\{R^{(1)}\}$, $E\{R^{(2)}\}$ and $E\{R^{(1)}\}-E\{R^{(2)}\}$ over a range of number of cycles of the type 1 marked job. Theoretical values for these quantities are shown in parentheses. Thus, for example, 100 cycles of the type 1 marked job were observed in the simulated time interval (0,903.00) and there were a total of 446 transitions in the continuous time Markov chain $\{L(t):t\geq 0\}$. A total of 130 response times for the type 1 (marked) job were observed along with 56 response times for the type 2 (marked) job. For the quantity $E\{R^{(1)}\}=7$, the point estimate 6.946 was obtained, and the 90 percent confidence interval had half-length 0.6334. Note that for $E\{R^{(1)}\}$ and $E\{R^{(2)}\}$, all of the confidence intervals surround the theoretical values. In the case of $E\{R^{(1)}\}-E\{R^{(2)}\}$, the confidence intervals based on Equation (10.2.8) also surround the theoretical value. Table 10.2 gives results obtained for $P\{R^{(1)}\leq x\}$, with $x=4$, 8, 12, 16 and 20.

In Table 10.3 we give, for the several values of x, point and interval estimates for $P\{R^{(1)}\le x\}-P\{R^{(2)}\le x\}$, based on the use of super-cycles and Equation (10.2.12). Thus, for $x=4$, 100 cycles based on response times for the type 1 job resulted in 37 super-cycles defined by response times for the type 2 job greater than x. Note that the number of cycles for the type 1 marked job has been fixed, and for each x the estimates for $P\{R^{(1)}\le x\}-P\{R^{(2)}\le x\}$ computed from the resulting random number of super-cycles.

Table 10.4 contains estimates of the quantities $P\{R^{(2)}\le x\}$ obtained from the standard regenerative method applied to these super-cycles. An overall observation from Tables 10.2 and 10.4 is that the lengths of confidence intervals obtained for $P\{R^{(1)}\le x\}$ and $P\{R^{(2)}\le x\}$ are roughly comparable.

TABLE 10.1

Simulation Results for Response Times in Closed Network of Queues With Two Job Types. $N_1=1$, $N_2=1$, $p=0.75$, $\lambda_1=1$, $\lambda_2=0.5$. Return State is (0,0,2,0,0,1,2,1,0,1,2,0,0,0,1,2).

	No. of Cycles For Type 1 Marked Job				
	100	200	400	800	1000
Total simulated time	903.00	1959.09	4134.11	8211.57	10172.83
No. of transitions/ cycle (M.C.)	4.46	4.85	5.18	5.12	5.15
No. of type 1 response times/cycle	1.30	1.40	1.48	1.46	1.47
No. of type 2 response times/cycle	0.56	0.65	0.75	0.74	0.74
$E\{R^{(1)}\}$ (=7)	6.946 ±0.6334	6.997 ±0.3931	6.995 ±0.2703	7.018 ±0.2033	6.920 ±0.1800
$E\{R^{(2)}\}$ (=14)	16.125 ±2.8713	15.187 ±1.8017	13.780 ±0.9851	13.965 ±0.7361	13.729 ±0.6305
$E\{R^{(1)}\}-E\{R^{(2)}\}$ (=-7)	-9.179 ±2.7421	-8.190 ±1.7316	-6.785 ±0.9145	-6.947 ±0.6809	-6.808 ±0.5848

TABLE 10.2

Percentiles of Type 1 Response Times in Closed Network of Queues With Two Job Types. $N_1=1$, $N_2=1$, $p=0.75$, $\lambda_1=1$, $\lambda_2=0.5$. Return State is (0,0,2,0,0,1,2,1,0,1,2,0,0,0,1,2).

	No. of Cycles For Type 1 Marked Job				
	100	200	400	800	1000
$P\{R^{(1)}\le 4\}$	0.2384 ±0.0622	0.2536 ±0.0417	0.2555 ±0.0301	0.2641 ±0.0217	0.2639 ±0.0192
$P\{R^{(1)}\le 8\}$	0.6692 ±0.0683	0.6714 ±0.0444	0.6717 ±0.0308	0.6709 ±0.0221	0.6802 ±0.0201
$P\{R^{(1)}\le 12\}$	0.8923 ±0.0422	0.8786 ±0.0293	0.8832 ±0.0205	0.8769 ±0.0159	0.8830 ±0.0140
$P\{R^{(1)}\le 16\}$	0.9461 ±0.0311	0.9536 ±0.0198	0.9594 ±0.0135	0.9547 ±0.0105	0.9605 ±0.0088
$P\{R^{(1)}\le 20\}$	0.9923 ±0.0127	0.9892 ±0.0100	0.9915 ±0.0061	0.9880 ±0.0052	0.9898 ±0.0043

TABLE 10.3

Difference of Percentiles of Response Times in Closed Network of Queues With Two Job Types. $N_1=1$, $N_2=1$, $p=0.75$, $\lambda_1=1$, $\lambda_2=0.5$. Return State is (0,0,2,0,0,1,2,1,0,1,2,0,0,0,1,2).

	No. of Cycles For Type 1 Marked Job				
	100	200	400	800	1000
$P\{R^{(1)}\le 4\}-P\{R^{(2)}\le 4\}$	0.1254 ±0.0784	0.1295 ±0.0627	0.1342 ±0.0417	0.1332 ±0.0275	0.1384 ±0.0192
No. of super-cycles	37	79	181	347	438
$P\{R^{(1)}\le 8\}-P\{R^{(2)}\le 8\}$	0.3988 ±0.1227	0.3409 ±0.0817	0.3111 ±0.0525	0.3036 ±0.0373	0.3125 ±0.0201
No. of super-cycles	29	61	134	254	319
$P\{R^{(1)}\le 12\}-P\{R^{(2)}\le 12\}$	0.3915 ±0.1424	0.3543 ±0.1112	0.3331 ±0.0677	0.3259 ±0.0496	0.3280 ±0.0140
No. of super-cycles	22	46	93	180	224
$P\{R^{(1)}\le 16\}-P\{R^{(2)}\le 16\}$	0.2850 ±0.1288	0.2970 ±0.1092	0.2693 ±0.0606	0.2693 ±0.0489	0.2636 ±0.0088
No. of super-cycles	15	32	65	129	153
$P\{R^{(1)}\le 20\}-P\{R^{(2)}\le 20\}$	0.2422 ±0.1081	0.2470 ±0.0825	0.2119 ±0.0530	0.2078 ±0.0415	0.2009 ±0.0043
No. of super-cycles	11	24	43	84	104

TABLE 10.4

Percentiles of Type 2 Response Times in Closed Networks of Queues With Two Job Types. $N_1=1$, $N_2=1$, $p=0.75$, $\lambda_1=1$, $\lambda_2=0.5$. Return State is (0,0,2,0,0,1,2,1,0,1,2,0,0,0,1,2).

	No. of Cycles For Type 1 Marked Job				
	100	200	400	800	1000
$P\{R^{(2)}\le 4\}$	0.1071 ±0.0852	0.1240 ±0.0583	0.1200 ±0.0312	0.1310 ±0.0209	0.1255 ±0.0183
No. of super-cycles	37	79	181	347	438
$P\{R^{(2)}\le 8\}$	0.2679 ±0.0963	0.3281 ±0.0751	0.3600 ±0.0475	0.3673 ±0.0323	0.3681 ±0.0280
No. of super-cycles	29	61	134	254	319
$P\{R^{(2)}\le 12\}$	0.5000 ±0.1184	0.5234 ±0.0799	0.5500 ±0.0474	0.5510 ±0.0336	0.5548 ±0.0286
No. of super-cycles	22	46	93	180	224
$P\{R^{(2)}\le 16\}$	0.6607 ±0.1155	0.6563 ±0.0687	0.6900 ±0.0419	0.6854 ±0.0316	0.6969 ±0.0268
No. of super-cycles	15	32	65	129	153
$P\{R^{(2)}\le 20\}$	0.7500 ±0.0986	0.7422 ±0.0582	0.7793 ±0.0405	0.7802 ±0.0284	0.7889 ±0.0243
No. of super-cycles	11	24	43	84	104

11.0. IMPLEMENTATION CONSIDERATIONS

In order to carry out a passage time simulation of a network of queues, we must be able to generate sample paths or realizations of the stochastic system. A necessary part of any such generation procedure is an algorithm (or algorithms) for random number generation, i.e., for the generation of numbers that can be treated as instances (samples) of random variables. In this section we consider aspects of random number generation pertinent to the implementation of passage time simulations according to the methods of the previous sections.

11.1. Random Number Generators

Our discussion follows Learmonth and Lewis (1973a). By a "random number generator" (or "pseudo-random number generator") we mean an algorithm which produces <u>sequences of numbers</u> that follow a specified <u>probability distribution</u> and possess the <u>appearance of randomness</u>. The use of "sequence of numbers" means that the algorithm is to produce many random numbers in a serial fashion. Even though a particular user may need only relatively few of the numbers, we generally require that the algorithm be capable of producing many numbers. "Probability distribution" implies that we can associate a probability statement with the occurrence of each number produced by the algorithm. We usually take the probability distribution to the uniform distribution on the interval [0,1]. If a source of [0,1] uniform random numbers is available, then in principle it is possible to transform these uniform random numbers by means of the inverse probability integral into random numbers having any desired distribution. For reasons of computational efficiency, however, a large amount of effort has gone into the development of methods for direct generation of random numbers having nonuniform distributions; see Ahrens and Dieter (1973a) for a comprehensive discussion. With respect to "appearance of randomness," it may be somewhat surprising that the actual implementation of most commonly used algorithms for uniform random number generation is as a (deterministic) recurrence relation in which each succeeding number is a function of the preceding number. Thus, although true randomness requires independence of successive numbers, the algorithm generates a deterministic dependent sequence. When parameters of the recurrence relation are chosen carefully, such algorithms for uniform random number generation do yield sequences which (statistically) appear to be random. This appearance of randomness is the origin of the term "pseudo-random numbers."

Since the results of a simulation depend critically on an acceptable appearance of randomness, it is important that a proposed uniform random number generator be subjected to thorough statistical testing. Although the simulation practitioner need not necessarily be concerned with the details of the rather specialized techniques for statistical testing of random number generators, he should be convinced prior to use that an available uniform random number generator has been successfully tested. See Fishman (1978), Ch. 8 for a discussion of statistical tests for uniform random number generators.

The most widely used (uniform) random number generators are of a class known as *linear congruential generators*. Such generators employ a recurrence relation of the form

$$X_n = bX_{n-1} + c \pmod{m} \tag{11.1.1}$$

In Equation (11.1.1) all quantities are nonnegative integers. This equation, read "X_n equals $bX_{n-1}+c$ modulo m," says that X_n is the remainder when $bX_{n-1}+c$ is divided by m. The quantity b is called the multiplier, m is the modulus, and c is the increment. Given a starting value $X_0 \geq 0$ and values $b \geq 0$, $c \geq 0$, and m such that $m > X_0$, $m > b$ and $m > c$, a sequence of integers $X_1, X_2, \ldots$ is generated by successive application of Equation (11.1.1). Uniform random numbers U_n on the interval [0,1] are obtained by dividing by m, i.e., for n=1,2,...,

$$U_n = X_n/m \tag{11.1.2}$$

The recurrence relation of Equation (11.1.1) is sometimes called a "mixed linear congruential generator," the term "mixed" coming from the fact that it involves a multiplication by a constant b along with an addition of a constant c. Many random number generators are "multiplicative" or "pure congruential" in that c=0, giving

$$X_n = bX_{n-1} \pmod{m} . \tag{11.1.3}$$

The initial or starting value X_0 is often called the seed of the random number generator.

Although it may appear that Equation (11.1.1) produces m distinct numbers, this is not the case unless b and m are chosen properly. It is characteristic of generators of this type that there is ultimately a cycle of numbers which is repeated indefinitely; this repeating cycle of numbers is called the period of the generator. It is clear that a congruential sequence used as a source of random numbers should have a long period, and since the period can never be greater than m, the value of m should be rather large.

Mathematical results based on number-theoretic considerations are available for characterizing the values of b and c which result in the maximum period length m; see Knuth (1969), Ch. 3. For the special case of multiplicative congruential generators (c=0), the basic result concerning maximum period length says that the maximum period length (m) is not achievable. It is, however, still possible to obtain multiplicative congruential generators with quite long periods. Results characterizing the maximum period for this multiplicative case are available, but the number-theoretic considerations are somewhat involved.

If the modulus m in a multiplicative congruential generator is prime, (i.e., has no divisors other than 1 and itself) a period of length m-1 is achievable. Such a period length, of course, is just one less than the maximum possible length. If, in addition, we choose the multiplier b so as to satisfy an appropriate (sufficient) number-theoretic condition with respect to (prime) m, then for any starting value $X_0<m$, the maximum period length m-1 is achieved. The determination of values for multipliers b satisfying the number-theoretic condition for maximum period length in a multiplicative congruential generator in general involves lengthy calculations. Further details are in Knuth (1969), Ch. 3.

In any particular digital computer system, only a finite number of positive integers are representable, the limitation being the word size of the system. We now state a particular (multiplicative congruential) uniform random number generator which utilizes the full word size of IBM System/360 (370) computer systems. (This is the uniform random generator used to obtain the numerical results in previous sections.) In the System/360, the word size is 32 bits with 1 bit reserved for algebraic sign; an obvious choice for m is thus 2^{31}. A multiplicative congruential generator with $m=2^k$ (for some positive integer k) can have a maximum period length of m/4. Thus for System/360 computer systems with $m=2^{31}$, the maximum period is 2^{29}, and the period length may also depend on the starting value. It happens (fortuitously) that the largest prime less than or equal to 2^{31} is $2^{31}-1$. Hence, by choosing $m=2^{31}-1$, it is possible to implement on System/360 computer systems uniform random number generators having a maximum period length of $2^{31}-2$. Note that the number-theoretic conditions ensuring a maximum period length do not necessarily guarantee good statistical properties for the generator, although the choice of the particular multiplier 7^5 does satisfy some known conditions regarding statistical properties of the resulting sequence.

System/360 Generator

Let $X_0>0$. Then for $n\geq 1$,

$$\begin{aligned} X_n &= 7^5\, X_{n-1} \pmod{2^{31}-1} \\ &= 16807\, X_{n-1} \pmod{2^{31}-1} \qquad (11.1.4) \end{aligned}$$

and

$$U_n = X_n/(2^{31}-1) \tag{11.1.5}$$

The uniform random number generator of Equation (11.1.5) has been tested extensively, and the results of the statistical tests indicate that it is very satisfactory; see Lewis, Goodman and Miller (1969) and Learmonth and Lewis (1973b). Other multipliers for generators with modulus $m=2^{31}-1$ are in use. Results of pertinent statistical tests are given by Hoaglin (1976).

11.2. Nonuniform Random Numbers

The problem of generating random numbers from a specified (nonuniform) distribution is in principle solved by having a source of uniform random numbers and transforming these random numbers by means of the inverse probability integral. Because it is not always possible to compute or to compute efficiently the inverse of a given distribution function, a great deal of effort has gone into the development of methods for direct generation of nonuniform random numbers; see Ahrens and Dieter (1973b) for a comprehensive discussion. Desirable properties of such direct methods are that they be exact, very fast, and economical of computer storage. The property of exactness is that any deviation from the specified distribution results from computer round-off error rather than a defect in the method itself. Comparisons are hard to make among particular methods, partly because of machine dependencies. It is, however, almost always true that with very little cost in complexity, it is possible to improve on the inverse probability integral transformation by an order of magnitude in execution time; the fastest available algorithms for nonuniform random number generation, however, are often quite complex.

The basis for the generation of nonuniform random numbers by transformation of a uniform random number is the following statement. If U is uniformly distributed on [0,1] and if F(x) is any distribution function, then the random variable

$$X = F^{-1}(U)$$

has distribution F(x); for $0 \le u \le 1$, the inverse function $F^{-1}(u)$ is defined by

$$F^{-1}(u) = \inf\{z : F(z) \ge u\} .$$

It follows that to generate samples of a random variable X having distribution F(x) from a random number U (uniformly distributed on [0,1]), we must be able to solve the equation

$$F(x) = u .$$

Then given a uniform random number U, we return

$$X = \inf\{z : F(z) \ge U\} .$$

Figure 11.1 illustrates the inverse transformation method. Note that this technique applies to discrete as well as continuous random variables.

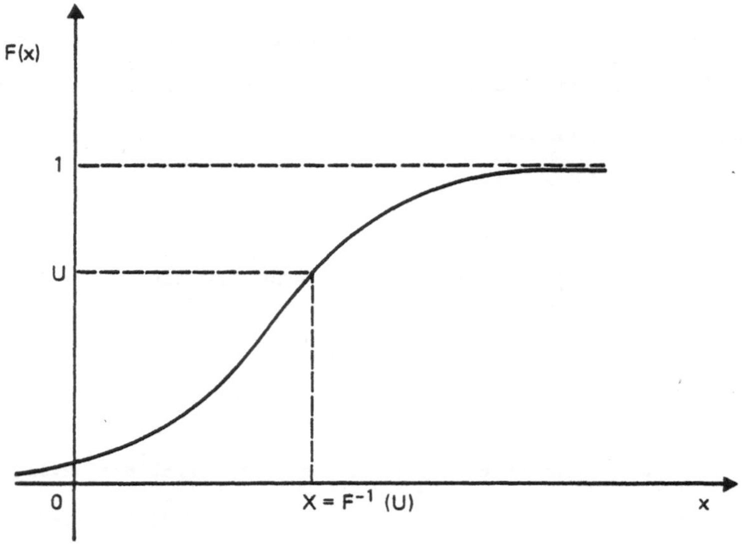

Figure 11.1. Inverse transformation method

The inverse transformation method provides a straightforward means of generating samples of an exponential random variable, as may be needed for a passage time simulation of a network of queues. We can obtain an exponential (rate parameter λ) random number X by generating U, a uniform random number on [0,1], and transforming it according to

$$X = -(\ell n\ U)/\lambda\ .$$

This transformation is obtained by solving the equation $U=F(X)$ for X, yielding $X=\{-\ell n(1-U)\}/\lambda$, and observing that $1-U$ is uniformly distributed on the interval [0,1] when U is. Note, however, that although this logarithmic transformation is easy to implement, it is a relatively slow method for obtaining exponential random numbers. The fastest methods available at present for exponential random numbers use so-called "decomposition" methods using the ideas of Marsaglia, MacLaren and Bray (1964). The basis for the method is division of the random variable into several populations, from most of which samples can be obtained easily. Using geometric considerations, the density function of the exponential random variable is decomposed into a large number of rectangular regions, wedge-shaped regions, and a tail. The Naval Postgraduate School random number generator package LLRANDOM (Learmonth and Lewis (1973a)) contains an IBM System/360 Basic Assembler Language implementation of a decomposition method for generation of exponential random numbers.

In implementation of a passage time simulation, it may be necessary to generate samples of a random variable X having a Cox-phase (exponential stage) representation. It is easy to do so if generators of uniform and exponential random numbers are availacle. Using the notation of Section 3.1, suppose that the distribution of X has n exponential stages. For $j=1,2,\ldots,n$,

$$X = X_1 + X_2 + \ldots + X_j$$

with probability $(1-b_j)a_j$, where the random variables $X_1,X_2,\ldots,X_j$ are mutually independent and X_i is exponentially distributed with rate parameter λ_i. To obtain random samples of X, we use a two step procedure. First, we generate U, a uniform random number on [0,1], and set k equal to 1 if $0<U\leq(1-b_1)a_1$ and equal to

$$\min\{j>1:(1-b_{j-1})a_{j-1}<U\leq(1-b_j)a_j\}$$

otherwise. Then, using the inverse transformation method for exponential random numbers, we generate k (mutually independent) uniform random numbers $U_1,U_2,\ldots,U_k$ and return

$$X = \sum_{j=1}^{k} (-\ln U_j)/\lambda_j \; .$$

The inverse transformation method for generating random numbers having a specified discrete distribution provides a means of routing jobs through a network of queues. In the simplest case (e.g., in the network of Figure 5.1), it is necessary to generate samples of a Bernoulli random variable W for which (with $0<p<1$)

$$P\{W=1\} = p$$

and

$$P\{W=0\} = 1-p \; .$$

To do so, we generate a random number U (uniformly distributed on [0,1]), and return W=1 if U≤p, and return W=0 otherwise. By this procedure, we are in effect partitioning the interval [0,1] and determining the value of X by the portion of the interval in which the generated uniform random number lies. The generalization to handle more complex routing from a service center is straightforward.

11.3. Single and Multiple Streams of Random Numbers

It is typically the case when simulating a network of queues that we have need for several streams of random numbers (e.g., random service times for each of several servers, routing through the network, etc.). Since most algorithms for nonuniform random number generation require the generation of one of more uniform random numbers, the question arises as to whether single or multiple streams of uniform random numbers should be used. When a single stream of uniform random numbers is used, the course of the simulation determines, usually in a complex manner, the role of individual uniform random numbers; thus, e.g., for the cyclic queues of Figure 2.1, a random subsequence of the generated uniform random numbers can be transformed to give the service times at one of the service centers, with the remaining random numbers used to generate the service times at the other center. Alternatively, if appropriate seeds are available, we can use nonoverlapping portions of the uniform random number sequence to generate the service times at the individual centers. The concern is that when a single stream of uniform random numbers is used, we are in effect assuming that particular random subsequences of the original uniform random number sequence has an acceptable appearance of randomness, and this may not be the case. There are examples of simulations where the use of a single stream of uniform random numbers has led to rather bizarre results. Although this aspect of random number generation is not well-understood, in many cases it is probably good practice to use separate streams. Some additional bookkeeping is of course necessary to handle the separate streams, and judgement is required as to what extent multiple streams should be used when simulating a complex stochastic system. For a network of queues, it is probably advisable to use separate random number streams for the interarrival times, service times, and routing of jobs from the individual service centers.

In Table 11.1 we give values of seeds which can be used to generate independent streams of uniform random numbers from the System/360 uniform random number generator of Equations (11.1.4) and (11.1.5). These seeds are values of X_n which are 100000 apart in the sequence of Equation (11.1.4), i.e., if X_0=377003613, then X_{100000}=648473574, X_{200000}=1396717869, etc. It is necessary when using multiple streams of random numbers to keep in mind approximately how many random numbers are needed; undesired dependence among random numbers may result if portions of the original sequence overlap inadvertently.

11.4. Generation of State Vector Processes

When carrying out a simulation of a network of queues (or any stochastic system), we observe the behavior of the system as it evolves in time. Implicit in any implementation of the simulation is the definition of an appropriate system state vector. This "state of the system at time t" constitutes a stochastic process, and to obtain estimates of quantities of interest, we must somehow generate realizations or sample paths of this state vector process. For complex networks of queues, it is often convenient to generate the process (e.g., using an event scheduling approach) by means of timing routines applicable to a general discrete event simulation, as typically provided by a high level simulation programming language. However, when there is a characterization of the state vector process as a familiar stochastic process, it may be possible to generate the process directly and more efficiently (with respect to speed) than by using all of the apparatus for timing which is necessary for a general discrete event simulation. This is relevant to the passage time simulations discussed here in that the state vector process $\{X(t):t\geq 0\}$ of Section 3.2 constitutes a finite state continuous time Markov chain. We consider generation of such Markov chains next.

Let $\underset{\sim}{Y}=\{Y(t):t\geq 0\}$ be a continuous time Markov chain having finite state space E, and let

$$\underset{\sim}{Q} = (q_{ij})$$

be its matrix of infinitesimal transition parameters; thus for $i,j\in E$, $q_{ij}\geq 0$ for $i\neq j$, and for all i,

$$\sum_{j\in E} q_{ij} = 0 \ .$$

Denote by $\{\tau_n:n\geq 0\}$ the jump times of the process, and for $n=0,1,\ldots$, set

$$Y_n = Y(\tau_n) \ .$$

Generation of the continuous time Markov chain $\underset{\sim}{Y}$ can be based on the following characterization (see, e.g., Çinlar (1975a), p. 247). For any $j \in E$, $u \in R_+$, and $n=0,1,\ldots$,

$$P\{Y_{n+1}=j,\ \tau_{n+1}-\tau_n>u \mid Y_0,\ldots,Y_n;\ \tau_0,\ldots,\tau_n\} = r_{ij}e^{-q_i u} \qquad (11.4.1)$$

if $X_n=i$. Here, $q_i=-q_{ii}$, $r_{ij}=q_{ij}/q_i$ for $i \neq j$, and $r_{ii}=0$. Thus, given a jump to state i, the process remains in state i for an exponentially distributed (rate parameter q_i) amount of time, and then jumps to state j with independent probability r_{ij}. It follows that generation of a sample path for the Markov chain $\underset{\sim}{Y}$, (i.e., the sequence of jump times and successive states) can be accomplished by successive generation of a pair of independent random numbers. This pair consists of an exponential random number and a sample from a discrete distribution specified by the jump probabilities. Note that each element of such a pair can be generated by the inverse transformation method.

This discussion of the generation of continuous time Markov chains presupposes that the elements of the $\underline{Q}$-matrix (and hence the jump probabilities) are available explicitly. For the class of networks of queues discussed here, enumeration of the state space and explicit calculation of the infinitesimal transition parameters is in general somewhat tedious. It is important to observe, however, that complete knowledge of the $\underline{Q}$-matrix (e.g., for the Markov chain defined by Equation (3.2.3)) is contained in the (given) routing matrix $\underline{P}$ and the parameters of the (exponential or Cox-phase) service times. Consequently, based on Equation (11.4.1), it is often possible to construct a more efficient algorithm for direct simulation of a given network of queues than that which results from a general discrete event simulation. For the DL/I component model of Section 5.2, Appendix 4 gives such an algorithm.

TABLE 11.1

Seeds for System/360 Uniform Random Number Generator.

Values of X_n 100,000 apart in $X_n = 7^5\ X_n-1$ (mod $2^{31}-1$).

(to be read across)

377003613	648473574	1396717879	2027350275	1356162430
1752629996	745806097	201331468	1393552473	1966641861
711072531	769795447	1074543187	1933483444	625102656
1116874679	1442211901	989455196	1996695068	1850124212
1267310126	1741371275	886499692	1014119573	933913228
2082204497	920168983	1079618777	1888797415	1002901030
1582733583	254293472	1095895189	219529399	1706847402
1951007719	1169002398	1482199345	1976077334	775245191
1976418161	35067978	400884188	1895732964	1904749580
1301700180	63685808	936615625	110322717	1029730003
251900732	725094089	828842333	1471230052	1703522097
1356420548	1670372925	437765009	39279049	2123613511
150006407	1633650593	751601611	1410990605	1262214427
645360044	1504645702	1063375004	941885586	1753135176
253642018	1701685042	1448665492	1034856864	428280431
259758456	600732272	704726097	398944698	114386769
288727775	1499601820	2136214308	1197972807	1888007825
686553263	747119178	154337000	136758808	9182540
303111010	154232008	921093990	1684263351	1166344707
1167753617	1374693082	1812641667	502455872	857532898

APPENDIX 1. CONVERGENCE OF PASSAGE TIMES

Label the jobs from 1 to N and for $t \geq 0$ set

$$Y(t) = (Z(t), N^1(t), N^2(t), \ldots, N^N(t)) ,$$

where $N^i(t)$ denotes the position in the job stack at time t of the job labelled i. The vector $Z(t)$ is the same as in Section 3.2. Also define the marginal processes

$$Y^i(t) = (Z(t), N^i(t))$$

for $i=1,2,\ldots,N$. All of the processes $\{Y(t): t \geq 0\}$ and $\{Y^i(t): t \geq 0\}$ are irreducible, positive recurrent continuous time Markov chains defined on a common underlying probability triple, $(\Omega, \mathscr{F}, P)$, say. Observe that if the marked job of Section 3 is the job labelled i, then the process $\{Y^i(t): t \geq 0\}$ coincides with the process $\{X(t): t \geq 0\}$ defined by Equation (3.2.3), except possibly for the initial condition at $t=0$. Define for each job two sequences of times, the starts and terminations of the successive passage times for the job. For the job labelled i, denote these times by $\{S_j^i: j \geq 0\}$ and $\{T_j^i: j \geq 1\}$. The definition of these times in terms of the process $\{Y^i(t): t \geq 0\}$ is completely analogous to what was done in Section 4.2 in terms of the process $\{X(t): t \geq 0\}$. Then the jth passage time for the job labelled i is $P_j^i = T_j^i - S_{j-1}^i$, $j \geq 1$. Also define Markov chains $\{X_j^i: j \geq 0\}$ for each job in which X_j^i denotes the state of the Markov chain $\{Y^i(t): t \geq 0\}$ when the $(j+1)^{st}$ passage time starts for job i: $X_j^i = Y^i(S_j^i)$. At this point we have N Markov renewal processes, $\{(X_j^i, S_j^i): j \geq 0\}$, all defined on $(\Omega, \mathscr{F}, P)$.

Next we introduce a new sequence of passage times, $\{P'_j:j\geq 1\}$, also defined on $(\Omega,\mathcal{F},P)$; this is the sequence of passage times irrespective of job identity, enumerated in order of start times. For each j, P'_j is a random member of the set $\{P^i_\ell:1\leq\ell\leq j\}$; this means that $P'_j=P^{\ell(j)}_{k(j)}$, where $\ell(j)$ and k(j) are random variables.

The principal result of this Appendix is to show that all of the sequences $\{P'_j:j\geq 1\}$ and $\{P^i_j:j\geq 1\}$ converge in distribution to a common random variable P.

(A1.1) PROPOSITION. For i=1,2,...,N, $P^i_j \Rightarrow P$ as $j\to\infty$. In addition, $P'_j \Rightarrow P$ as $j\to\infty$.

Proof. Since the N jobs are identical with respect to their service requirements and branching probabilities, the semi-Markov kernels governing the Markov renewal processes $\{(X^i_j,S^i_j):j\geq 0\}$ all coincide with the kernel $\underset{\sim}{K}$ of Section 4.2. In fact, for any particular job the only difference from the setup of Section 4.2 is that (with possibly one exception) the job does not start a passage time at t=0. However, this difference does not alter limiting results; the job labelled i starts a passage time with probability one (since $\{Y^i(t):t\geq 0\}$ is positive recurrent), and once this occurs, the situation is exactly as in Section 4.2. Note in particular that $S^i_j\to+\infty$ a.s. for all i; thus, there is always a next passage time for every job. This being so, we have $P^i_j \Rightarrow P$ as $j\to\infty$, just as was the case in Section 4.2. (This result is to be expected since the marked job was selected arbitrarily.)

Next we show that $P'_j \Rightarrow P$. Since $P^i_j \Rightarrow P$ for all i, we can use the Skorohod representation theorem (see Skorohod (1956) or Billingsley (1971)) to assert the existence of a probability space $(\tilde{\Omega},\tilde{\mathcal{F}},\tilde{P})$ and random variables $\tilde{P}^i_j$ ($j\geq 1$, $1\leq i\leq N$) and $\tilde{P}$ defined on that space such that $\tilde{P}^i_j$ and $\tilde{P}$ have the same distributions as P^i_j and P, respectively, and $\tilde{P}^i_j \to \tilde{P}$ a.s. as $j\to\infty$ for all i. These representatives $\tilde{P}^i_j$ also provide representatives for the P'_j which we call $\tilde{P}'_j$.

Putting aside the null sets of $\tilde{\Omega}$ on which the above convergence statements do not hold, we examine the numerical sequence $\{\tilde{P}'_j(\omega):j\geq 1\}$ for one of the remaining $\omega\in\tilde{\Omega}$. We use the following criterion for convergence of a numerical sequence $\{x_i:j\geq 1\}$: $x_j\to x$ as $j\to\infty$ if and only if for each subsequence $\{x_{j'}\}$ there exists a further subsequence $\{x_{j''}\}$ that converges to x; see Billingsley (1968), p. 15, for a similar usage in weak convergence theory. Select a subsequence $\{\tilde{P}'_{j'}(\omega)\}$. This subsequence must contain a further subsequence $\{\tilde{P}'_{j''}(\omega)\}$ that is identical to a subsequence of one of the sequences $\{\tilde{P}^i_{j'}(\omega):j\geq 1\}$, say for $i=i_0$. This follows from the fact that there are only a finite number of jobs. But this subsequence $\{\tilde{P}^{i_0}_{j'}\}$ converges a.s. to $\tilde{P}(\omega)$ since the full sequence does. Thus $\tilde{P}'_j \to \tilde{P}$ a.s. and therefore $P'_j \Rightarrow P$.

APPENDIX 2. PROOF OF RATIO FORMULA

We provide a proof of the ratio formula (Theorem (4.2.9)) for $E\{f(X,P)\}$ which makes it possible to use the regenerative method for estimation of passage times. The proof given here does not require the key renewal theorem.

(A2.1) PROPOSITION. Assume that $E\{|f(X,P)|\}<\infty$. Then $E\{f(X,P)\}=E\{Y_1(f)\}/E\{\alpha_1\}$, where $Y_1(f)$ is given by Equation (4.2.6).

<u>Proof</u>. Assume $f\geq 0$, $E\{f(X,P)\}<\infty$, and set $f_c=\min(f,c)$ for some c such that $0<c<\infty$. Clearly, Equation (4.2.8) also holds for f_c, i.e.,

$$f_c(X_n,P_{n+1}) \Rightarrow f_c(X,P) ,$$

and since f_c is bounded,

$$\lim_{n\to\infty} E\{f_c(X_n,P_{n+1})\} = E\{f_c(X,P)\} . \tag{A2.2}$$

Next we compute the Cesáro average of the sequence appearing in Equation (A2.2). First we write

$$E\left\{\sum_{n=0}^{m} f_c(X_n,P_{n+1})\right\}/(m+1) = E\left\{\sum_{k=1}^{\ell(m)+1} Y_k(f_c)\right\}/(m+1) - E\{Y'(m)\}/(m+1) , \tag{A2.3}$$

where $\ell(m)=\max\{k:\beta_k\leq m\}$ and $Y'(m)=\sum_{n=m+1}^{\beta_{\ell(m)+1}} f_c(X_n,P_{n+1})$.

Since $0\leq f_c\leq c$, we have $0\leq Y'(m)\leq c(\beta_{\ell(m)+1}-m)$. In addition, Wald's equation (see Chung (1968), Theorem 5.5.3) implies that

$$E\{\beta_{\ell(m)+1}\} = E\{\alpha_1\}E\{\ell(m)+1\}$$

and

$$E\left\{\sum_{k=1}^{\ell(m)+1} Y_k(f_c)\right\} = E\{Y_1(f_c)\}E\{\ell(m)+1\} \ .$$

These equations plus the elementary renewal theorem (Smith (1958), p. 246) imply that

$$\lim_{m\to\infty} E\{\beta_{\ell(m)+1}-m\}/(m+1) = 0$$

and

$$\lim_{m\to\infty} E\left\{\sum_{k=1}^{\ell(m)+1} Y_k(f_c)\right\} /(m+1) = E\{Y_1(f_c)\}/E\{\alpha_1\} \ .$$

Hence from Equation (A2.3) we have

$$\lim_{m\to\infty} E\left\{\sum_{n=0}^{m} f_c(X_n,P_{n+1})\right\} /(m+1) = E\{Y_1(f_c)\}/E\{\alpha_1\} \ . \tag{A2.4}$$

From Equations (A2.2) and (A2.4) we conclude that

$$E\{f_c(X,P)\} = E\{Y_1(f_c)\}/E\{\alpha_1\} \ . \tag{A2.5}$$

Now we let $c\to\infty$ on both sides of Equation (A2.5) and use the assumption that $E\{f(X,P)\}<\infty$ to obtain

$$E\{f(X,P)\} = E\{Y_1(f)\}/E\{\alpha_1\} \ . \tag{A2.6}$$

For a general f function, we write $f=f^+-f^-$ and apply the above argument to both f^+ and f^-. Thus we have Equation (A2.6) provided $E\{|f(X,P)|\}<\infty$.

APPENDIX 3. ESTIMATION OF VARIANCE CONSTANTS

We first consider estimation of the variance constant σ^2 appearing in Equation (10.2.8) which leads to a confidence interval for $E\{R^{(1)}\}-E\{R^{(2)}\}$. Based on n cycles, for i=1,2, compute $\hat{\sigma}_{ii}$ as an estimate of

$$\sigma_{ii} = E\{(Z_k^{(i)})^2\} = \text{var}\{\alpha_k\} - 2r^{(i)}\text{cov}\{\alpha_k,N_k^{(1)}\} + (r^{(i)})^2\text{var}\{N_k^{(1)}\}$$

according to

$$\hat{\sigma}_{ii} = s_{11} - 2\hat{r}_n^{(i)}s_{12}^{(i)} + (\hat{r}_n^{(i)})^2 s_{22}^{(i)} ,$$

where

$$s_{11} = (n-1)^{-1} \sum_{j=1}^{n}(\alpha_j-\bar{\alpha}_n)^2 ,$$

$$s_{12}^{(i)} = (n-1)^{-1} \sum_{j=1}^{n}(\alpha_j-\bar{\alpha}_n)(N_j^{(i)}-\bar{N}_n^{(i)}) ,$$

and

$$s_{22}^{(i)} = (n-1)^{-1} \sum_{j=1}^{n}(N_j^{(i)}-\bar{N}_n^{(i)})^2 ,$$

with

$$\bar{\alpha}_n = n^{-1} \sum_{j=1}^{n}\alpha_j,\ \bar{N}_n^{(i)} = n^{-1} \sum_{j=1}^{n}N_j^{(i)},\ \text{and}\ \hat{r}_n^{(i)} = \bar{\alpha}_n/\bar{N}_n^{(i)} .$$

Finally, compute $\hat{\sigma}_{12}$ as an estimate of

$$\sigma_{12} = \text{var}\{\alpha_k\} - r^{(1)}\text{cov}\{\alpha_k,N_k^{(1)}\} - r^{(2)}\text{cov}\{\alpha_k,N_k^{(2)}\} + r^{(1)}r^{(2)}\text{cov}\{N_k^{(1)},N_k^{(2)}\}$$

according to

$$\hat{\sigma}_{12} = s_{11} - \hat{r}_n^{(1)}s_{12}^{(1)} - \hat{r}_n^{(2)}s_{12}^{(2)} + \hat{r}_n^{(1)}\hat{r}_n^{(2)}s_{22} ,$$

where s_{11}, $s_{12}^{(1)}$ and $s_{12}^{(2)}$ are as before, and

$$s_{22} = (n-1)^{-1} \sum_{j=1}^{n} (N_j^{(1)} - \bar{N}_n^{(1)})(N_j^{(2)} - \bar{N}_n^{(2)}) .$$

Then estimate σ^2 according to

$$\hat{\sigma}^2 = \frac{\hat{\sigma}_{11}}{(\bar{N}_n^{(1)})^2} + \frac{\hat{\sigma}_{22}}{(\bar{N}_n^{(2)})^2} - \frac{2\hat{\sigma}_{12}}{\bar{N}_n^{(1)}\bar{N}_n^{(2)}} .$$

In an analogous manner, we estimate the variance constant $\sigma^2(x)$ appearing in Equation (10.2.12) which leads to a confidence interval for $P\{R^{(1)} \le x\} - P\{R^{(2)} \le x\}$. Based on n super-cycles, for $i=1,2$ compute $\hat{\sigma}_{ii}(x)$ as an estimate of

$$\sigma_{ii}(x) = \mathrm{var}\{Y_k^{(i)}\} - 2P\{R^{(i)} \le x\}\mathrm{cov}\{Y_k^{(i)}, n_k^{(i)}\}$$
$$+ (P\{R^{(i)} \le x\})^2 \mathrm{var}\{n_k^{(i)}\}$$

according to

$$\hat{\sigma}_{ii}(x) = s_{11}^{(i)}(x) - 2\left(\frac{\bar{Y}_n^{(i)}}{\bar{n}_n^{(i)}}\right) s_{12}^{(i)}(x) + \left(\frac{\bar{Y}_n^{(i)}}{\bar{n}_n^{(i)}}\right)^2 s_{22}^{(i)}(x)$$

where

$$s_{11}^{(i)}(x) = (n-1)^{-1} \sum_{j=1}^{n} (Y_j^{(i)} - \bar{Y}_n^{(i)})^2 ,$$

$$s_{12}^{(i)}(x) = (n-1)^{-1} \sum_{j=1}^{n} (Y_j^{(i)} - \bar{Y}_n^{(i)})(n_j^{(i)} - \bar{n}_n^{(i)}) ,$$

and

$$s_{22}^{(i)}(x) = (n-1)^{-1} \sum_{j=1}^{n} (n_j^{(i)} - \bar{n}_n^{(i)})^2 ,$$

with

$$\overline{Y}_n^{(i)} = n^{-1} \sum_{j=1}^{n} Y_j^{(i)} \text{ and } \overline{n}_n^{(i)} = n^{-1} \sum_{j=1}^{n} n_j^{(i)} .$$

Finally, compute $\hat{\sigma}_{12}(x)$ as an estimate of

$$\begin{aligned}\sigma_{12}(x) = \; & \mathrm{cov}\{Y_k^{(1)},Y_k^{(2)}\} - P\{R^{(1)}\le x\}\mathrm{cov}\{Y_k^{(2)},n_k^{(1)}\} \\ & - P\{R^{(2)}\le x\}\mathrm{cov}\{Y_k^{(1)},n_k^{(2)}\} \\ & + P\{R^{(1)}\le x\}P\{R^{(2)}\le x\}\mathrm{cov}\{n_k^{(1)},n_k^{(2)}\}\end{aligned}$$

according to

$$\begin{aligned}\hat{\sigma}_{12}(x) = \; & s_{11}^{(1)}(x) - \left(\frac{\overline{Y}_n^{(1)}}{\overline{n}_n^{(1)}}\right) s_{12}^{(1)}(x) - \left(\frac{\overline{Y}_n^{(2)}}{\overline{n}_n^{(2)}}\right) s_{21}^{(2)}(x) \\ & + \left(\frac{\overline{Y}_n^{(1)}\overline{Y}_n^{(2)}}{\overline{n}_n^{(1)}\overline{n}_n^{(2)}}\right) s_{22}(x) ,\end{aligned}$$

where $s_{11}^{(1)}(x)$, $s_{22}^{(1)}(x)$, and $s_{12}^{(2)}(x)$ are as before, and

$$s_{22}(x) = (n-1)^{-1} \sum_{j=1}^{n} (n_j^{(1)}-\overline{n}_n^{(1)})(n_j^{(2)}-\overline{n}_n^{(2)}) .$$

Then estimate $\sigma^2(x)$ according to

$$\hat{\sigma}^2(x) = \frac{\hat{\sigma}_{11}(x)}{(\overline{n}_n^{(1)})^2} + \frac{\hat{\sigma}_{22}(x)}{(\overline{n}_n^{(2)})^2} - \frac{2\hat{\sigma}_{22}(x)}{\overline{n}_n^{(1)}\overline{n}_n^{(2)}} .$$

APPENDIX 4. GENERATION OF MARKOV CHAIN IN DL/I COMPONENT MODEL

For $1 \le i \le 7$, denote by λ_i the rate parameter of the exponentially distributed service time for jobs of class i. Complete knowledge of the $\underline{Q}$-matrix of infinitesimal transition parameters for the continuous time Markov chain $\underset{\sim}{X}=\{X(t):t\ge 0\}$ defined by Equations (5.2.1) and (5.2.2) is contained in the routing matrix $\underline{P}$ and the λ_i. We give an algorithm for direct generation of this process.

1. Fix an initial state in A_2 and set $t=0$.

2. Determine the status of the α and β center servers, i.e., whether or not they are busy, and if so, what classes of jobs are in service. The β center server is busy if $Q(t)>0$; the α center server is busy if $S(t)>0$.

Assume both are busy, with a class $i>0$ job in service at the α center. The cases in which only one server is busy are handled similarly.

3. Generate a holding time τ, exponentially distributed with rate parameter $\lambda_1+\lambda_i$ and advance t to $t'=t+\tau$.

4. Determine according to a Bernoulli trial the service which completes first:

$$P\{\alpha \text{ center service completes first}\} = \frac{\lambda_i}{\lambda_1+\lambda_i}$$

and

$$P\{\beta \text{ center service completes first}\} = \frac{\lambda_2}{\lambda_1+\lambda_i}\ .$$

If α center service completes first, go to 6.

5. Set

$$Q(t') = Q(t)-1 ,$$

$$C_5(t') = C_5(t)+1 ,$$

$$S(t') = S(t) ,$$

and for $2 \le j \le 7$ and $j \ne i$, set

$$C_j(t') = C_j(t) .$$

If $S(t') \ne 7$, go to 7. Otherwise, set

$$C_7(t') = C_7(t)+1 ,$$

$$C_5(t') = C_5(t')-1 ,$$

$$S(t') = 5 ,$$

and go to 7 .

6. For $2 \le k \le 7$, set

$$C_k(t') = C_k(t) .$$

Generate $i \in \{1,2,\ldots,7\}$ according to the probabilities p_{ji}, where $S(t)=j$, and set

$$C_i(t') = C_i(t')+1, \; i>1$$

and

$$Q(t') = Q(t)+1, \; i=1 .$$

Set

$$S(t') = k_0$$

and

$$C_{k_0}(t') = C_{k_0}(t')-1 ,$$

where $k_0 = \min\{k : C_k(t')>0\}$.

7. Determine $N(t')$ as follows. If the job completing service is the marked job, set $N(t')$ equal to its position in the job stack

after completion of the service. If the job completing service is not the marked job, set N(t')=N(t) if the job completing service goes back in the job stack either above [respectively below] the marked job, when prior to completion the job was above [respectively below] the marked job. Set N(t')=N(t)-1 if job completing service goes from above the marked job to below. Otherwise, set N(t')=N(t)+1.

8. Return to 2. and iterate with t' playing the role of t.

REFERENCES

Ahrens, J. H. and Dieter, U. (1973a). Uniform Random Numbers. Institut für Mathematische Statistik. Technische Hochschule in Graz. Graz, Austria.

Ahrens, J. H. and Dieter, U. (1973b). Nonuniform Random Numbers. Institut für Mathematische Statistik. Technische Hochschule in Graz. Graz, Austria.

Billingsley, P. (1968). Convergence of Probability Measures. John Wiley. New York.

Billingsley, P. (1971). Weak Convergence of Measures: Applications in Probability. Society of Industrial and Applied Mathematics. Philadelphia, Pennsylvania.

Chung, K. L. (1967). Markov Chains with Stationary Transition Probabilities. Springer-Verlag. Berlin.

Chung, K. L. (1968). A Course in Probability Theory. Harcourt, Brace, and World. New York.

Çinlar, E. (1975a). Introduction to Stochastic Processes. Prentice-Hall. Englewood Cliffs, New Jersey.

Çinlar, E. (1975b). Markov renewal theory: A survey. Management Sci. 21, 727-752.

Crane, M. A. and Iglehart, D. L. (1975a). Simulating stable stochastic systems: III, Regenerative processes and discrete event simulation. Operations Res. 23, 33-45.

Crane, M. A. and Iglehart, D. L. (1975b). Simulating stable stochastic systems, IV: Approximation techniques. Management Sci. 21, 1215-1224.

Crane, M. A. and Lemoine, A. J. (1977). An Introduction to the Regenerative Method for Simulation Analysis. Lecture Notes in Control and Information Sciences, Vol. 4. Springer-Verlag. Berlin Heidelberg New York.

Cox, D. R. (1955). A use of complex probabilities in the theory of stochastic processes. Proc. Cambridge Philos. Soc. 51, 313-319.

Cox, D. R. and Lewis, P. A. W. (1966). The Statistical Analysis of Series of Events. Methuen. London. Distributed by Barnes and Noble. New York.

Feller, W. (1968). An Introduction to Probability Theory and Its Applications. Vol. I (3rd Ed.). Wiley. New York.

Fishman, G. S. (1973). Statistical analysis of queueing simulations. Management Sci. 20, 363-369.

Fishman, G. S. (1978). Principles of Discrete Event Simulation. Wiley. New York.

Gelenbe, E. and Muntz, R. R. (1976). Probabilistic models of computer systems - Part I (Exact results). Acta Informat. 7, 35-60.

Heidelberger, P. (1977). Variance reduction techniques for the simulation of Markov processes, I: Multiple estimates. Technical Report No. 42. Department of Operations Research. Stanford University. Stanford, California.

Hoaglin, D. (1976). Theoretical properties of congruential random-number generators: An empirical view. Memorandum NS-340. Department of Statistics. Harvard University. Cambridge, Massachusetts.

Hordijk, A., Iglehart, D. L. and Schassberger, R. (1976). Discrete time methods for simulating continuous time Markov chains. Advances in Appl. Probability 8, 772-778.

Iglehart, D. L. (1975). Simulating stable stochastic systems, V: Comparison of ratio estimators. Naval Res. Logist. Quart. 22, 553-565.

Iglehart, D. L. (1976). Simulating stable stochastic systems, VI: Quantile estimation. J. Assoc. Comput. Mach. 23, 347-360.

Iglehart, D. L. (1978). The regenerative method for simulation analysis. Current Trends in Programming Methodology Vol. III: Software Engineering. K. M. Chandy and R. T. Yeh (eds.), 52-71. Prentice-Hall. Englewood Cliffs, New Jersey.

Iglehart, D. L. and Shedler, G. S. (1978a). Regenerative simulation of response times in networks of queues. J. Assoc. Comput. Mach. 25, 449-461.

Iglehart, D. L. and Shedler, G. S. (1978b). Simulation of response times in finite capacity open networks of queues. Operations Res. 26, 896-914.

Iglehart, D. L. and Shedler, G. S. (1979a). Regenerative simulation of response times in networks of queues with multiple job types. Acta Informat. 12, 159-175.

Iglehart, D. L and Shedler, G. S. (1979b). Regenerative simulation of response times in networks of queues: statistical efficiency. IBM Research Report RJ 2587. San Jose, California.

Iglehart, D. L. and Shedler, G. S. (1979c). Simulation methods for response times in networks of queues. IBM Research Report RJ 2628. San Jose, California. To be presented at 1979 Winter Simulation Conference.

Knuth, D. L. (1969). The Art of Computer Programming. Vol. 2, Semi-Numerical Algorithms. Addison-Wesley. Reading, Massachusetts.

Lavenberg, S. S. and Sauer, C. H. (1977). Sequential stopping rules for the regenerative method of simulation. IBM J. Res. Develop. 21, 545-558.

Lavenberg, S. S. and Shedler, G. S. (1976). Stochastic modelling of processor scheduling with application to data base management systems. IBM J. Res. Develop. 20, 437-448.

Lavenberg, S. S. and Slutz, D. R. (1975). Regenerative simulation of a queuing model of automated tape library. IBM J. Res. Develop. 19, 463-475.

Learmonth, G. P. and Lewis, P. A. W. (1973a). Naval Postgraduate School random number package LLRANDOM. Naval Postgraduate School Report NPS55Lw73061A. Monterey, California.

Learmonth, G. P. and Lewis, P. A. W. (1973b). Statistical tests of some widely used and recently proposed uniform random number generators. Proc. Conference on Computer Science and Statistics: 7th Annual Symposium on the Interface. Iowa State University. Western Publishing Co. No. Hollywood, California, 163-171.

Lewis, P. A. W. (1964). A branching Poisson process model for the analysis of computer failure patterns. J. Roy. Statist. Soc. Ser. B 26, 398-451.

Lewis, P. A. W., Goodman, A. S., and Miller, J. M. (1969). A pseudo-random number generator for the System/360. IBM Systems J. 8, 199-220.

Lewis, P. A. W. and Shedler, G. S. (1971). A cyclic-queue model of system overhead in multiprogrammed systems. J. Assoc. Comput. Mach. 18, 199-220.

Marsaglia, G., MacLaren, M. D. and Bray, T. A. (1964). A fast procedure for generating normal random variables. Comm. Assoc. Comput. Mach. 7, 4-10.

Miller, D. R. (1972). Existence of limits in regenerative processes. Ann. Math. Statist. 43, 1275-1282.

Reiser, M. and Sauer, C. H. (1978). Queuing network models: methods of solution and their program implementation. Current Trends in Programming Methodology Vol. III: Software Engineering. K. M. Chandy and R. T. Yeh (eds.), 115-167. Prentice-Hall. Englewood Cliffs, New Jersey.

Shedler, G. S. (1979). Regenerative simulation of response times in networks of queues, III: Passage through subnetworks. IBM Research Report RJ 2466. San Jose, California.

Skorohod, A. V. (1956). Limit theorems for stochastic processes. Theor. Probability Appl. 1, 262-290 (English translation).

Smith, W. L. (1958). Renewal theory and its ramifications. J. Roy. Statist. Soc. Ser. B 20, 243-302.

INDEX OF NOTATION

Lecture Notes in Economics and Mathematical Systems

For information about Vols. 1–114 please contact your bookseller or Springer-Verlag

Vol. 115: Anderson et al., Foundations of System Theory: Finitary and Infinitary Conditions. VII, 93 pages. 1976

Vol. 116: K. Miyazawa, Input-Output Analysis and the Structure of Income Distribution. IX, 135 pages. 1976.

Vol. 117: Optimization and Operations Research. Proceedings 1975. Edited by W. Oettli and K. Ritter. IV, 316 pages. 1976.

Vol. 118: Traffic Equilibrium Methods, Proceedings 1974. Edited by M. A. Florian. XXIII, 432 pages. 1976.

Vol. 119: Inflation in Small Countries. Proceedings 1974. Edited by H. Frisch. VI, 356 pages. 1976.

Vol. 120: G. Hasenkamp, Specification and Estimation of Multiple-Output Production Functions. VII, 151 pages. 1976.

Vol. 121: J. W. Cohen, On Regenerative Processes in Queueing Theory. IX, 93 pages. 1976.

Vol. 122: M. S. Bazaraa, and C. M. Shetty,Foundations of Optimization VI. 193 pages. 1976

Vol. 123: Multiple Criteria Decision Making. Kyoto 1975. Edited by M. Zeleny. XXVII, 345 pages. 1976.

Vol. 124: M. J. Todd. The Computation of Fixed Points and Applications. VII, 129 pages. 1976.

Vol. 125: Karl C. Mosler. Optimale Transportnetze. Zur Bestimmung ihres kostengünstigsten Standorts bei gegebener Nachfrage. VI, 142 Seiten. 1976.

Vol. 126: Energy, Regional Science and Public Policy. Energy and Environment I. Proceedings 1975. Edited by M. Chatterji and P. Van Rompuy. VIII, 316 pages. 1976.

Vol. 127: Environment, Regional Science and Interregional Modeling. Energy and Environment II. Proceedings 1975. Edited by M. Chatterji and P. Van Rompuy. IX, 211 pages. 1976.

Vol. 128: Integer Programming and Related Areas. A Classified Bibliography. Edited by C. Kastning. XII, 495 pages. 1976.

Vol. 129: H.-J. Lüthi, Komplementaritäts- und Fixpunktalgorithmen in der mathematischen Programmierung. Spieltheorie und Ökonomie. VII, 145 Seiten. 1976.

Vol. 130: Multiple Criteria Decision Making, Jouy-en-Josas, France. Proceedings 1975. Edited by H. Thiriez and S. Zionts. VI, 409 pages. 1976.

Vol. 131: Mathematical Systems Theory. Proccedings 1975. Edited by G. Marchesini and S. K. Mitter. X, 408 pages. 1976.

Vol. 132: U. H. Funke, Mathematical Models in Marketing. A Collection of Abstracts. XX, 514 pages. 1976.

Vol. 133: Warsaw Fall Seminars in Mathematical Economics 1975. Edited by M. W. Loś, J. Loś, and A. Wieczorek. V. 159 pages. 1976.

Vol. 134: Computing Methods in Applied Sciences and Engineering. Proceedings 1975. VIII, 390 pages. 1976.

Vol. 135: H. Haga, A Disequilibrium – Equilibrium Model with Money and Bonds. A Keynesian – Walrasian Synthesis. VI, 119 pages. 1976.

Vol. 136: E. Kofler und G. Menges, Entscheidungen bei unvollständiger Information. XII, 357 Seiten 1976.

Vol. 137: R. Wets, Grundlagen Konvexer Optimierung. VI, 146 Seiten. 1976.

Vol. 138: K. Okuguchi, Expectations and Stability in Oligopoly Models. VI, 103 pages. 1976.

Vol. 139: Production Theory and Its Applications. Proceedings. Edited by H. Albach and G. Bergendahl. VIII, 193 pages. 1977.

Vol. 140: W. Eichhorn and J. Voeller, Theory of the Price Index. Fisher's Test Approach and Generalizations. VII, 95 pages. 1976.

Vol. 141: Mathematical Economics and Game Theory. Essays in Honor of Oskar Morgenstern. Edited by R. Henn and O. Moeschlin. XIV, 703 pages. 1977.

Vol. 142: J. S. Lane, On Optimal Population Paths. V, 123 pages. 1977.

Vol. 143: B. Näslund, An Analysis of Economic Size Distributions. XV, 100 pages. 1977.

Vol. 144: Convex Analysis and Its Applications. Proceedings 1976. Edited by A. Auslender. VI, 219 pages. 1977.

Vol. 145: J. Rosenmüller, Extreme Games and Their Solutions. IV, 126 pages. 1977

Vol. 146: In Search of Economic Indicators. Edited by W. H. Strigel. XVI, 198 pages. 1977.

Vol. 147: Resource Allocation and Division of Space. Proceedings. Edited by T. Fujii and R. Sato. VIII, 184 pages. 1977.

Vol. 148: C. E. Mandl, Simulationstechnik und Simulationsmodelle in den Sozial- und Wirtschaftswissenschaften. IX, 173 Seiten. 1977.

Vol. 149: Stationäre und schrumpfende Bevölkerungen: Demographisches Null- und Negativwachstum in Österreich. Herausgegeben von G. Feichtinger. VI, 262 Seiten. 1977.

Vol. 150: Bauer et al., Supercritical Wing Sections III. VI, 179 pages. 1977.

Vol. 151: C. A. Schneeweiß, Inventory-Production Theory. VI, 116 pages. 1977.

Vol. 152: Kirsch et al., Notwendige Optimalitätsbedingungen und ihre Anwendung. VI, 157 Seiten. 1978.

Vol. 153: Kombinatorische Entscheidungsprobleme: Methoden und Anwendungen. Herausgegeben von T. M. Liebling und M. Rössler. VIII, 206 Seiten. 1978.

Vol. 154: Problems and Instruments of Business Cycle Analysis. Proceedings 1977. Edited by W. H. Strigel. VI, 442 pages. 1978.

Vol. 155: Multiple Criteria Problem Solving. Proceedings 1977. Edited by S. Zionts. VIII, 567 pages. 1978.

Vol. 156: B. Näslund and B. Sellstedt, Neo-Ricardian Theory. With Applications to Some Current Economic Problems. VI, 165 pages. 1978.

Vol. 157: Optimization and Operations Research. Proceedings 1977. Edited by R. Henn, B. Korte, and W. Oettli. VI, 270 pages. 1978.

Vol. 158: L. J. Cherene, Set Valued Dynamical Systems and Economic Flow. VIII, 83 pages. 1978.

Vol. 159: Some Aspects of the Foundations of General Equilibrium Theory: The Posthumous Papers of Peter J. Kalman. Edited by J. Green. VI, 167 pages. 1978.

Vol. 160: Integer Programming and Related Areas. A Classified Bibliography. Edited by D. Hausmann. XIV, 314 pages. 1978.

Vol. 161: M. J. Beckmann, Rank in Organizations. VIII, 164 pages. 1978.

Vol. 162: Recent Developments in Variable Structure Systems, Economics and Biology. Proceedings 1977. Edited by R. R. Mohler and A. Ruberti. VI, 326 pages. 1978.

Vol. 163: G. Fandel, Optimale Entscheidungen in Organisationen. VI, 143 Seiten. 1979.

Vol. 164: C.-L. Hwang and A. S. M. Masud, Multiple Objective Decision Making – Methods and Applications. XII, 351 pages. 1979.

Vol. 165: A. Maravall, Identification in Dynamic Shock-Error Models. VIII, 158 pages. 1979.

Vol. 166: R. A. Cuninghame-Green, Minimax Algebra. XI, 258 pages. 1979.

Vol. 167: M. Faber, Introduction to Modern Austrian Capital Theory. X, 196 pages. 1979.

Vol. 168: Convex Analysis and Mathematical Economics. Proceedings 1978. Edited by J. Kriens. V, 136 pages. 1979.

Vol. 169: Coalition Formation by Sophisticated Players. By A. Rapoport et al. VII, 170 pages. 1979.